The Art of Music Publishing

*FROM THOSE THAT HAVE BEEN
GIVEN THE MOST,
MUCH IS EXPECTED.*

The Art of Music Publishing

An entrepreneurial guide to publishing and copyright for the music, film and media industries

Helen Gammons

ELSEVIER

AMSTERDAM • BOSTON • HEIDELBERG • LONDON • NEW YORK • OXFORD
PARIS • SAN DIEGO • SAN FRANCISCO • SYDNEY • TOKYO

Focal Press is an imprint of Elsevier

Focal Press is an imprint of Elsevier
The Boulevard, Langford Lane, Kidlington, Oxford, OX5 1GB
30 Corporate Drive, Suite 400, Burlington, MA 01803, USA

First published 2011

British Library Cataloguing in Publication Data
A catalogue record for this book is available from the British Library

Library of Congress Number: 2010939822

ISBN: 978-0-240-52235-7

For information on all Focal Press publications
visit our website at www.focalpress.com

Printed and bound in the UK

10 11 12 13 12   11 10 9 8 7 6 5 4 3 2 1

Working together to grow
libraries in developing countries

www.elsevier.com | www.bookaid.org | www.sabre.org

ELSEVIER   BOOK AID International   Sabre Foundation

Contents

Introduction — How it all began

This book is dedicated to two wonderful friends and business mentors; two people with hearts and spirits that truly inspired me. They made life fun and definitely amusing.

To Chris Brough, who worked with artists such as Cat Stevens, Shakin' Stevens, Samantha Fox, Showaddywaddy and Jaki Graham, to name but a few. Having substantial UK and international success, Chris was a producer, music publisher, manager and visionary, who lived life to the full. His success helped him to build an incredible network, which he generously shared over the many years that we knew each other.

And to Mike Collier, ex-MD of Carlin/Intersong Music, who was an inspirational mentor and friend. He signed such acts as Duran Duran, and the disco hit 'Blame It On The Boogie', made famous by the legendary Michael Jackson with a top 3 and top 8 in the US and UK charts, respectively, and several million record sales and millions of radio performance plays later. Hits continued with Phyllis Nelson's song 'Move Closer' becoming a worldwide hit having been covered by countless artists since.

Mike's personal perseverance (despite ill-health in later years) never stopped his strong belief in the value of songs and music. Mike was a 'can do' man and I hope this characteristic and unstoppable optimism radiates from me too. In his later years, Mike worked as catalogue activator and compilation guru alongside Freddie Faber, and they were infamous in London for knowing the established major publishers' catalogues like the backs of their hands. At this time, mergers and acquisitions were rife, and as a consequence staff churn was high and this left a void where staff didn't know what they had in the catalogue. Individuals like Mike and Freddie knew where the gold was buried, and were retained by many of the major publishing houses to assist them in activating their back catalogue. This knowledge has largely been lost to younger publishing A&R personnel. It's still a problem today with all the mergers and acquisitions that have occurred (perhaps there is an even bigger need now than ever before where intellectual property management is the key to unlocking new revenue flows).

Mike was affectionately known as 'Grandpa disco' because of his knack for finding great pop tunes and turning classics into contemporary versions in those days. Later he became known as 'Grandpa line dance' when he single-handedly brokered many new deals for the use of masters and songs for line dancing videos and programs. Mike's breadth of legal expertise and his love and passion for music, and all things publishing, provided me with the essential training ground that has helped all aspects of my career ever since. He was the epitome of creative publishing.

I also dedicate this book to Mike's lovely wife Joyce and dearest friend Jack Robinson, whose catalogue Robin Song Music was administered by Mike for many years.

The night I met Mike is indelibly imprinted on my memory. Arriving at a dinner with my writer and producer husband Rod Gammons, we were met by our host, the

CEO of Record Shack Records, whose artists included Barbara Pennington, Hi NRG and Village People. We were escorted to the table where Mike took it upon himself to welcome guests, and on greeting me for the first time launched into an extremely cheeky and rather saucy comment, which had an immediate effect on the entire group of diners, who all burst out laughing. I would gladly share the actual words he uttered but his comment was far too rude to repeat here. Having to think on my feet, and fast, I managed to bat back an equally naughty reply. The table of guests became a complete hysterical mass of tears and laughter and it took some time for order to be resumed. It was to set the scene for an extremely enjoyable evening that would continue very late into the night, while starting a new friendship that would be sealed forever. Mike was totally incorrigible.

The year was 1982 and up until this time, I had managed and acted as a UK agent for local rock bands, supplying support acts to such major touring groups as Whitesnake, Black Sabbath and Hawkwind. However, my first big break came as a freelance radio plugger.

Alan Smith at Dakota Records was just coming off a No. 1 UK hit record with 'Zambeezee' and his radio plugger (Oliver Smallman) had been unable to secure any interest in a new act he was developing. Unperturbed and gloriously unaware of who Oliver Smallman was (he was the most successful radio plugger at that time), I arrived in London to plug my first record. Andy Kowan Martin (one of only two people I knew in London who worked in the industry) — manager of Bad Manners, and Kip Trevor (music publisher) met me in the Green Man pub just around the corner from Radio 1. They were so proud that I had secured my first job in the industry, but when I announced I was setting out to go and plug the single at Radio 1 that very day, and without any appointments prearranged, Andy literally fell off his stool with laughter. I was a little upset by this lack of faith in my ability. I completely finished him off when I asked him where Radio 1 was!

About 10 minutes later, when the tears of laughter had stopped rolling down his face and he could string a sentence together he finally composed himself and both he and Kip (having necked two large vodka and oranges) escorted me to the revolving door at Radio 1. (All these years later it still makes me laugh just thinking about that day and just how naive I was.)

Ah, the brave of heart. I was so determined I could have stopped an express train with a single hand! Arriving at the reception desk at Radio 1, I spent a difficult five minutes trying every angle possible to get the receptionist to give me some appointments. Pity had no place in her life. She was good … I had no impact on the ice queen at all! However, lurking behind me was a handsome young man who, upon hearing my pitiful excuses, came to my rescue. I had no idea who he was but, excited about his invitation to be escorted around Radio 1, followed him to the lift. I remember thinking, this is just too good to be true; Andy and Kip will never believe my luck!

As we walked into the lift I walked past a picture hung elegantly on the wall adjacent to the lift door, which portrayed Radio 1's director of programs, Chris

Lycett. Oh my god … this was who I was stepping into the lift with! I felt my heart grind to a halt as a flush of acute embarrassment washed over me. To say I was the color of a beetroot would have been an understatement.

With nowhere to hide within this tiny three-foot-square lift and no hole that I could find to swallow me up, I graciously thanked him repeatedly for his generosity. In the 20 seconds it took the lift to reach its destination I knew I had to rise to the challenge and seize this very special moment. Putting my nerves back into my very new leather briefcase I exited the lift as a well rounded, ready for anything, music industry professional. The transformation was immediate!

Chris was terribly kind and supportive and full of encouragement for my first commercial endeavors, and my plucky route into the industry. Not only did he personally take me into the belly of Radio 1, but he introduced me to every single DJ/presenter and personal assistant behind the scenes. I thought my pounding heart was loud enough to be transmitted over the airways. I enthusiastically presented the product to each person from Bruno Brooks to John Peel, focusing in on the key selling points of the product, the success of the label and the artist.

Two hours later I had met everyone, had had coffee with the PAs and staff and had been through the entire building. Thanks to Chris the record was played the following weekend on the new 'Famous Five' slot; and most importantly I got paid by Dakota Records. I was a success! I had money to stay alive a few more months, pay my rent and work towards my next job. I earnestly developed my network of business contacts, and tried to piece together just how it was that the industry worked.

Everyone I met seemed to know a little piece of the jigsaw puzzle but few understood how the industry worked as a whole. The music industry was like a dark art, a secret club; no one really knew much beyond their own little world. I was hooked and I knew I wanted to know everything, and this was to be my journey; a journey that I am still on today.

Radio promotion — exciting as it is — was not to be my ultimate calling, although I did continue with a few records after this. It was during this time that I met Tony ('premo') Peters, who was MD at Acuff Rose in the UK. A wonderfully enthusiastic 'old school' publisher, he often toured the bowels of Radio 1, saying hello to people and pushing songs that he published (which had been recorded by various artists and released by a variety of different labels). In the 1980s and 1990s it was common to have music publishers doing as much work as the record pluggers in promoting records at radio. The music publisher wanted to receive maximum plays, while the pluggers and labels wanted the highest chart position from sales, which meant that radio was essential. Both had their ways and methods of working radio. Relationships were, and still are, key and who you know is an essential ingredient alongside a great record.

It was around this time when I first realized the financial impact of having a hit song and what this meant to the music publisher and their revenue stream. A song could literally be reborn every time an artist recorded it! How cool is that? It creates a brand new income stream from that version of the song, and new money for the writers and publishers. It suddenly became very exciting to me. Instantly I saw that while an artist might have a very short-lived and precarious career, a song could have many chances of becoming successful, and that great songs were successful time and time again! Just take a look at the charts: on albums of all genres, covers are everywhere!

The artist was the vehicle, and a great artist made the song their own; but a great song had many owners, crossed genres, fashion, even decades, to find relevance and a place in the hearts of so many different people. The world of music publishing was, and still is, the most fascinating area of the music and media industry. I'm still in love after all these years.

Having only a limited knowledge of the business side of the industry at that time, it was to Mike Collier that I turned for advice in putting our first publishing company, Seaside Music, on the international map. The route we took was to put in place a network of subpublishers all around the world who would (in their respective territory) assist us in both collecting royalties earned, and creating local covers in their market. At my second MIDEM conference[1] in 1987, I charged Mike with setting in place meetings with all the top MDs and CEOs of major and independent publishers around the world — I aimed for the top and got it in one. Our roles in these meetings were quite simple: I went in and made the sales pitch regarding the product and catalogue development. Mike negotiated the deal structure, advances and terms, and closed the deals there and then on a handshake. At that MIDEM we walked out having struck deals worth £125k in advances and had set in place the international structure through which we would trade (all based on music and songs). It was a truly exciting partnership that worked well for us both. By going country by country we could select the strongest partners to work with. It created more administration, but there were also added financial upsides. No territory was cross-recouped from another (more on this in Chapter 6).

However, it was my thirst for knowledge and understanding of the complexities of the industry that really fascinated me. Music publishing was the area of the industry that really excited me, but I wanted to understand the entire structure. Mike would continue to be my mentor for several years, until the day came when he actually turned around and asked me a question!

When this happened, Mike smiled and said: "My work is done, and the student outperforms the teacher". Mike was a music publisher and lawyer, having taken his bar exams in the USA. He was an immensely successful publisher, so much so that Freddie Bienstock apparently fought against paying him commission on classic titles he'd found, such as 'Blame It On The Boogie' (written not by Michael Jackson, star and legend, but by a UK writer, Mick Jackson), and artists such as Duran Duran. Mike was eventually triumphant. It was at this time that Mike went freelance and became a music consultant. His depth of knowledge, not only on matters of publishing but also on the songs and masters within these incredible catalogues, became a valuable asset to many companies including EMI, Music Sales, CBS/Sony

---

[1] The premier international music industry conference, held in Cannes, France, every year.

and many others with whom he was engaged as a back catalogue specialist consultant.

In the UK he administered Robin Song Music for his long-time friend Jack Robinson — the first international Robin Song Music hit was by Gloria Gaynor, 'If You Want It, Do It Yourself'. This was immediately followed up by England's Tina Charles with 'I Love To Love', which went to No. 1 in 19 countries and sold upwards of 20 million copies. Tina was in the Top 10 again in the UK with 'Love Me Like A Lover' and 'Rendezvous', which were also penned by Jack Robinson and his long-time partner David (James Bolden) Christie. Jack became a co-writer with my husband on several new David Christie titles and still has a successful company based in France.

Preface — change brings new opportunity

The music industry has been rocked by music piracy, incompetent business leaders, an inability to think strategically, an inability to put the customer and the music fan first, archaic accounting procedures, contracts that alienate artists, and music publishers that have used complex international accounting practices to reduce royalties to writers. Yet it has survived and it has begun to change; it is a business facing the future with new hope, with many more competent business leaders at its helm, with new, innovative and more transparent accounting practices being developed. It is fighting for its survival at a time when new opportunities come to market weekly. There are many things wrong with the industry, but there are many people fighting to put it right and to bring it safely into a more prosperous period so that creatives (artists, songwriter/composers, producers) continue to enrich our lives. Life without music is incomprehensible and so artists and composers and those who invest in them must be able to live and further invest in the future of our industry.

The question to ask is not: 'Is there a future for me in an industry with declining record sales?', but 'How can I take part in the new income streams that are emerging?'

Flexibility and agility in business are more essential now than ever before. There's money in your music, you've just got to know how to monetize it and collect it.

It is said that knowledge is power, well if not power, it certainly provides the foundation for informed decision making. How do you become better informed? Read! What should you read? There is a rich resource of information from industry books and weekly magazines to RSS feeds from top technology and media sites. These will be industry-wide resources and not specific to music publishing. When you understand that any changes in the industry ripple throughout it then you can appreciate how intertwined all facets of the industry are.

The most important thing is to stay informed and involved in the debate. I would say that the most consistent weakness I come across in young people is that for some reason they have stopped reading. But it's the easiest thing to put right. It's never too late to turn this weakness into a strength. Being well read and informed means you can take part in discussions from a position of knowledge. It improves your confidence and communication skills. Where should you start?

Wow — there's a lot out there! Reading *Music Week* (the UK industry music paper) is one of the best ways of seeing change in action. *Billboard* is the equivalent in the USA. Read industry-related reports from the BPI, ASCAP, Harry Fox, the International Federation of the Phonographic Industry (IFPI), BMI, PRS for Music, UK Music and of course any new political report; Gowers, Digital Britain (Digital Economy Bill), Media reports from Music Ally http://musically.com. Get involved in organizations such as the Music Managers Forum. Attend music conferences (local and international) and get involved in the industry at every level possible. Key industry stories and market changes are also reported in the more serious business newspapers and, of course, the wider business environment helps you to see the music industry in relationship to other industries. Market pressures will affect all businesses in some way. The music industry may have other specific pressures, but it too will be affected by the economic downturn, the lack of financial liquidity and stability, and the increasing aversion to risk. Change will inevitably mean some things will cease, and new things will replace them. The industry has coped well in the past with change, perhaps less so more recently. Change is exciting and frightening at the same time. It brings into view the role of technology and our role in applying it to business, the industry and a global market. It can lead to new opportunities and exciting new businesses and there will be winners and losers. But we must look to our fundamental principles to guide us. I believe that every rights owner has the right to be paid if they choose to write and/or perform music for a living, in the same way as anyone going to work expects to be paid.

Some writers can no longer afford to work full time as writers (and I'm talking about fairly established writers). As a society we must decide what role we want technology to play, what new business models we can embrace and how we can put the customer first, and yet reward and protect product creators effectively. Just remember, all these new business models take content (music and masters) and sell access or services on the back of this. They get rich while the copyright creators become worse off. In retail clothing, coffee production and all sorts of areas there is now a move to 'fair trade' business, and we must not exploit copyright creators and must pay due care and attention to their being able to sustain a living. This has to work hand in hand with strong business skills of meeting customer demand and need, and creating customer demand and need.

I'll leave with a question to ponder on, one I'm sure will take a few years to research. Are the diminishing returns to the industry purely down to music piracy (and this is what we are fighting), or do changing social attitudes and changing business practices also have an impact, and if so how much? Consider for a moment that Amazon sells second-hand books (I don't mind buying second hand); they also sell used CDs. Why would people buy used CDs and not download or pirate? But they do in their droves. What impact is this secondary market having?

How we change social attitudes may have a lot more to do with school and education in general. What work is being done by the music industry in schools worldwide to change the curriculum? Topics such as songwriting and performing, income streams, copyright, value and the right and wrong of using unlicensed download sites need to be covered.

CHANGE — A SWEEPING VIEW OVER THE CENTURIES

In the thirteenth century, putting a piece of music to paper meant manual scribing of the notes and often the score was embellished with hand decorations. This formed an important part of the monastic way of life. Moving forward, we pass the development of woodcut printing and then movable type, leading on to engraving, lithography and photolithography and then chromolithography (color).[2] While the technological changes were taking place the first copyright legislation came into force in 1710, with the 'Statute of Anne' granting exclusive rights to authors.[3] This was a key turning point and the bedrock to what is now our music publishing industry.[4]

If we whisk through another hundred years of change we pass the rise and fall of the importance of sheet music, the piano roll, the era of the silent movies, the impact of the talking movies and the vehicle of picture and music together and all the marketing initiatives that resulted. The development of radio brought music out of the back waters and into the towns and cities.[5] The reach of radio broadcasts, alongside changes in the commercial development of flight and communications, provided a landscape for labels and publishers to think beyond their shores with regard to selling records, and bringing artists new audiences. The music industry had become international and with every new technological development the industry found new ways to bring music to the masses and sell its products. We've had vinyl (starting with the old 78s, which are very brittle and major collectors' items these days), modern flexible vinyl, cassettes, VHS, Beta, CDs, mini disks, DVDs, Blu-rays, digital downloads, streaming and cloud. Coping with changing delivery platforms is not a new phenomenon to music publishers; they are used to being paid from a variety of income streams, licensing new uses and platforms, and we have the capacity to collect that money in several ways. Some income streams are royalty based, blanket fee based, buy-out fee based, percentage based, timing based and/or audience based.

The issue now is not what will come next, as change is inevitable, but the speed with which the industry can react to change. We need to work with and provide measured support to help new businesses to develop new concepts and in doing so develop new income streams for the industry. If the industry prevaricates too much we will see such government interference and pressure such as with the Gower report. The industry has had an attitude of 'stand in the queue and we'll get to you in time'. It must seriously embrace change and act in a manner that is far more proactive and entrepreneurial if it is to avoid seeing more and more of its controls and powerbase ebb away and our current allies and customers become competitors or owners.

Embrace change, but be ready for it!

---

[2]http://parlorsongs.com/insearch/printing/printing.php

[3]http://digital-law-online.info/patry/patry2.html

[4]http://www.copyrighthistory.com/anne.html

[5]http://www.peermusic.com/aboutus/companyhistory.cfm

HOW TO COMMUNICATE WITH NEW CUSTOMERS (NOT JUST CONSUMERS)

One must be proactive and authoritative in such a climate. If there's money to be made, let's get on and make it. The turmoil in the industry is often felt because the industry is too slow to react, not proactive enough and unwilling to understand what customers (in the broadest sense) want. For the industry to be better able to cope, it has to be far more entrepreneurial in characteristic, agile, strategic, visionary, and with a willingness to embrace change and engage with the consumer and the customer. This does not automatically mean compromised income, or a breakdown of copyright and the protection it affords or indeed accepting every new idea. Protecting copyright and assets is important.

Coping with change also requires the music industry to have the right people running its organizations. Without entrepreneurial traits at the heart of what we do, I fear that endeavors will be too slow, too little, too late, to cope in a fast, innovative, creative industry. By way of example, PRS for Music could be considered one of the best collection societies in the world but still tends to be slow moving in decision making. In contrast, HMV is a retailer with a vision and healthy diversification plans, as shared with the audience at June 2010's Great Escape conference in Brighton, UK.

UK Music's impact is clear as a strategic force for the future. Feargal Sharkey (legendary artist and songwriter) adds entrepreneurial traits to problem solving and has launched his 'Liberating Creativity' manifesto with a comprehensive battle plan. Being passionate about all that we do has to be at the heart of our actions, and Feargal is an exemplary illustration of this.

In 2008, the value of the UK music industry was £3.6 billion, as stated by PRS for Music's chief economist Will Page, a rise of 4.7% on the year before. The main area of growth was in business-to-business (B2B) including PPL income. The report also cites that digital income and non-traditional sources could bridge the losses in physical income by 2011, which is supported by some of the figures coming from the IFPI report 2010, looking at trading figures for 2009. The report supports my views that it is key that businesses are agile, innovative and visionary, and engage with the consumer.[6] PRS for Music goes on to state that there was a 2.6% rise in annual revenues to £623 million. Significant increases in revenues from British music used abroad (up 19.4% to £166.9 million) were buoyed by both exchange-rate gains and increased licensing activity in new and established territories. Online revenues grew by 72.7% to £30.4 million, reflecting the increased number of legal licensed digital music services available in the UK and across Europe. This growth (£12.8 million) outperformed the decline in traditional CD and DVD formats (down £8.7 million) for the first time, although the legal online music market is still comparatively small. Public performance revenues increased by 2.4% despite a reduction in licence fees

---

[6]http://www.prsformusic.com/creators/news/research/Documents/Will%20Page%20and%20Chris%20Carey%20(2009)%20Adding%20Up%20The%20Music%20Industry%20for%202008.pdf

for small businesses to £44 per annum, as more businesses took greater advantage of the benefits of music.

A SNAPSHOT OF THE IFPI REPORT 2010

- **Piracy** — is now causing the collapse of some local industries. In France local new artist signings fell by 59% on their level in 2002. In Spain local artist album sales in the Top 50 fell by 65% between 2004 and 2009. Brazil shows similar data.
- **Legislation** — France, New Zealand, South Korea, Taiwan and the UK have adopted or are proposing new measures requiring ISPs to tackle mass copyright theft on their networks.
- ISPs stand to gain from partnerships and see revenues in excess of 100 million.
- **Global music market** — fell by 7.2% in 2009. However, within this figure, growth was seen in 13 countries, hidden by the huge downturn in the USA and Japan.

RECORDED MUSIC SALES 2009

- Global recorded music trade revenues totaled US $17 billion in 2009, a decline of 7.2% on 2008. Physical sales continue to fall (−12.7%), a slowing decline from −15%.
- The sales base has shifted from physical to digital revenues; the latter grew by 9.2% in 2009 to US $4.3 billion (more than 10 times the digital market in 2004). Digital channels now account for 25.3% of all music sales.
- Performing rights revenues (digital) also grew strongly by 7.6% to US $0.8 billion. This reflects an unbroken trend of growth since 2003. Revenues from the sector now represent 4.6% of the total recorded music industry.
- Six markets (the UK, India, South Korea, Thailand, Mexico and Australia) experienced increased digital sales that exceeded the amount of physical decline: a landmark point.
- This is significant for the UK market: if this is achievable in times of recession then there can be true optimism moving forwards.

With over 30 years' experience in the music industry and armed with an MBA from Manchester Business School, I hope to share my experiences and inspire others, as I too have been inspired. My wish is to impart a knowledge that is based not just on sound business practices but also on hands-on experience.

We will journey together and set in motion entrepreneurial seeds that I hope will create a desire in you to find your own path and succeed in something you are passionate about. Don't sit on the fence: have a voice, get involved. It is a book to empower you, to 'think outside the box'; to be unconstrained in one's approach to ideas, and at the same time to assist you in using strategic business tools (that assist both the survival and growth of any business). These are especially important in enabling both the entrepreneur and the business manager to navigate any changing landscape. Whether you are a student, employed or self-employed this book is for you. Something here, I am sure, will be a catalyst in the weeks, months or years ahead, so keep in touch online.

With the landscape of the music industry changing more rapidly than ever, it will be an incredible journey! I realize as I finish this book that there will be changes that don't make it to print, but such is life. But we can discuss these via alternative communication routes. I will show you how to learn, adapt and develop no matter what the market is doing.

Get involved in predicting the new landscape that you will work in. Listen to those who are in the middle of these changes, and find out what they see and predict. Shape the future. One thing is for sure, change is constant, but the rate of change varies, and at the moment change is very fast.

Go and explore all the companies and websites I talk about.

Entrepreneurs are emerging across all areas of media. Look around you: even in the middle of a worldwide recession there are new opportunities opening up in both niche and macro environments. Above all, understand that the music industry is a business, and beset by pressures as are all businesses.

Changes in consumer behavior, market trends, economic pressures, the flow of money, technological advances and empowerment of the consumer affect all businesses including ours. The bar is being set higher and higher by companies pushing the envelope, and competition is fierce.

I will introduce you to industry leaders who were there at the beginning of some of the most revolutionary times in the music industry and for whom this new landscape is just as exciting. You will, I hope, learn to appreciate change, and not to be frightened of it. Change heralds a new dawn, the taking and development of new ideas and new possibilities by new people, and it signals the end of those ways that cannot continue, and therefore people and businesses that cannot adapt to change will fade. This does not mean anarchy, but evolution. This is why creators and those who manage them must have a strong voice.

Technology is providing the oxygen that is fueling the speed at which change is taking place. The questions that still need to be asked are: where is the market heading, and what might my customer want from me in the future? These are constant questions for all businesses, not just the music industry, and not just for today but for tomorrow. Determine who your customer is and for everyone reading this book the answer will be slightly different. The music industry has to date (in many respects) failed to do this. How many businesses can survive if they sue their own customer base (internet positioning/digital rights management/peer to peer)? No wonder there is a lack of customer loyalty and trust. Those businesses that are winning embrace the consumer and the customer, but again this does not mean pandering to consumers and becoming the victim.

As fresh opportunities open up, develop new networks and business partners in a variety of markets. Identify your approach to the Business-to-Consumer (B2C) and the B2B markets. Be aware of social, cultural and ethical issues that can affect how a business operates in different countries. As Seymour Stein said while attending Musexpo 2007[7] with me and 200 other executives: "I am here because I want to learn about doing business in China and Asia. You never stop learning in this business".

The music industry is a business and therefore conforms to key business analysis and structure, while also being tempered by the unpredictable element of 'creativity', often refusing to be constrained by regimental form. Yet it is in harnessing this creativity that new ideas and income streams can be developed, by the A&R person, the composer, the artist or the music publisher. It is the welding together of business and vision that has brought us to this point.

Only time will tell if Guy Hands and the EMI/Terra Firma deal will produce new growth (and not just a trimming of costs and perhaps improved bottom line). If he has a vision it is in putting the customer first (not last) and monetizing the true value of what they have; sitting down and doing business with all those companies to which the industry has traditionally been closed. Latest reports appear to show that perhaps the first good decision has been made and that has been to appoint a music publisher over both the record label and music publishing divisions. Music publishers understand 'rights management' and have no problem dealing with such matters alongside creative development work. The problem with most record labels is the need for volumetric releases to keep the war machine oiled. Perhaps the next step would be to separate out subdivisions within labels, allowing for business agility and faster decision making while achieving the economies of scale that come from being within the EMI Group. This would allow EMI to see the wood for the trees and bring costs under control.

We (the industry) have to affect the way change takes place, not react to it. In 2001, I made a recommendation as part of my Master's thesis and presentation to the 'rights' organizations at that time, that since the music industry was too fragmented to cope with 'industry unifying issues' it needed to formulate a cohesive strategy while tackling industry.

In 2008, UK Music was born and, finally, Feargal Sharkey[8] has provided a voice. He is not an administrator, he's a doer, and there are few people around who are prepared to take on this type of role. On meeting him recently at the Academy of Contemporary Music in Guildford (UK), I found his passion, persistence and vision equal to mine in providing young people with an inspiring and prepared route into the industry. I wish him well in all that he strives to do. UK Music's 'Liberating Creativity' manifesto is very well timed in the marketplace and with a change of government now in place this could do well.

Those individuals who I grew up with and who have remained in the industry have found a way through entrepreneurial enterprise to survive and grow, to combine

---

[7]A music industry conference held in Los Angeles in April for senior level executives. A good chance to network and combine the conference with meetings in LA. I use this to keep developing my networks at TV and film companies as well as with music supervisors.

[8]http://www.nme.com/artists/feargal-sharkey

a strong business discipline with passion and excitement. I count myself very lucky to still be working and earning my living from an industry that I fell in love with 30 years ago. In recent times I have found incredible satisfaction from helping young people to start their careers in the industry. Through various lecture work opportunities and mentoring programs, I have found another satisfying career path, in addition to running my own very successful group of companies with my husband and son. Being nominated for family business of the year in the middle of a recession is no small achievement and, to my team, well done!

A life without music is incomprehensible: society would be bereft of culture if this should happen and it won't, so don't listen to those promoting doom and gloom. The one thing that will survive is the song, and therefore the songwriter and performer will find an outlet. But as a society we must make sure that this talent is not lost, or the quality of such talent diminished in any way because they can no longer make a living wage. We must not rob our society of culture, and look back in sadness in 50 years' time and see a neutered landscape and musical innovation choked. Music is everywhere, and we can't live without it. As Quincy Jones said: "Music is the only product that is consumed before it is bought".[9] How right he is.

There is something here for everyone; from those just beginning in music publishing to those who perhaps would welcome a fresh perspective. For media companies wanting to do business with the music industry this book may also assist in the vocabulary used and the structure of today's industry. In an industry where a high proportion of individuals are entrepreneurs with little or no formal business education, it is a very powerful tool with which to harness both entrepreneurial traits and business acumen. For those looking for a career in the industry, it provides a great framework to enable you to think about and find future employment. For those looking to start their own business, it provides plenty of strategic ideas and tools to work with. Intellectual property (IP) rights are assets that the banks and even the City understand. Music publishing offers investors real returns, even these days. The same cannot be said of record labels.

So who invests in music and media projects? Let's take a quick look at a recent article:[10]

Icebreaker is an exciting investment opportunity in the licensing of Intellectual Property Rights.[11]

Ingenious[12]
Venture capital firm, Ingenious Media, announced in 2006 the launch of a new fund which will have millions to invest in small media companies, including those in the music industry. As of June 2010 the venture capital company continues to develop.

Power Amp Music Fund
Ex-Citi private banker launches innovative music fund

---

[9]http://www.quincyjones.com
[10]http://www.wealth-bulletin.com/home/content/1050293397

A tax-efficient investment fund could point to the future for the music industry.[13]

Tom Bywater, who quit his job as a discretionary portfolio manager at Citigroup two years ago, aims to raise £10m for the Power Amp Music Fund, structured as a UK Enterprise Investment Scheme. The money will be used to invest in between 20 to 30 acts, and produce and release their recordings, and promote tours and merchandising. "Unlike traditional recording contracts signed by the top labels, artists will not get big advance payments but will enjoy a much greater share of revenues", said Bywater in 2008. However, stop press March 2010 sees Charlotte Church agree a deal for £2 million on 50% of gross revenues from all Rights for a limited period. A rather significant advance in my opinion.[14]

Polyphonic Investment Fund

MAMA Group, Nettwerk Music Group and ATC announce the formation of artist investment business Polyphonic (July 2009)[15]

Three of the world's foremost artist management companies announced today a significant evolution in their engagement with the artist community.

MAMA Group, Nettwerk Music Group and ATC have announced the formation of Polyphonic, which will invest directly into artist businesses, thereby offering an alternative to the traditional label-driven investment model of the music industry.

Founded with an initial financial commitment of more than $20 million for its first year of operation (and with significant additional capital available thereafter), Polyphonic will seek to partner with artists and their business managers to provide the capital to enable them to operate their own businesses, retain their own copyrights and take a fair share of any profits that are generated.

Polyphonic, through a simple partnership governed by contract rather than corporate structure, will supply the investment required to enable an artist to build their business beyond the constraints of traditional business models. All copyrights will remain the property of the artist. Polyphonic will earn a share of the profits generated by all revenue earned from artist activities – in much the same way that artist's managers currently earn profits.

Pledge Music

Pledge Music, as reported by the *Guardian* newspaper, is an investment platform that offers fans the chance to support artists by becoming part of the record-making process.

"When Napster came on the scene, it felt like a big wave coming at us, and it appeared futile to try and stop it", says Benji Rogers, founder of London-based Pledge Music. Rogers, who has been working as a musician for more than a decade, has spent much of his time building a company that aims to surpass the fan-funding models of SliceThePie, Sellaband and Bandstocks. "We don't see the fans as venture capitalists, who are investing in our records", he says. Instead, Rogers wants them to feel part of the process that gets a record into their hands.

Established acts like Marillion, Public Enemy and Tina Dico (one of the first artists to use Pledge Music, raising £60,000 in just 30 days), have relied on their long-standing fan base when leaving a major label to pursue their own direction. But, can fan funding really work for artists who are just starting out? Rogers thinks so, but it takes time and effort.

The first question Pledge asks when approached by an artist is how big is their mailing list? Based on that, they calculate how much money they can raise. Once they've worked it out, the Pledge team sit down with the artist to find out what they can add to the fans' experience to raise that money. The higher the amount a fan pledges, the more 'extras' they get.[16]

[11]http://www.icebreakerfund.co.uk/home.asp

[12]http://www.ingeniousmedia.co.uk/asset-management/ingenious-funds/uk-equity-fund/the-ingenious-uk-equity-fund

[13]http://www.wealth-bulletin.com/home/content/1050293397

[14]http://musically.com/blog/2010/03/10/charlotte-church-takes-2m-investment-from-power-amp-music/

[15]http://www.nettwerk.com/blog/terry/mama-group-nettwerk-music-group-and-atc-announce-formation-artist-investment-business-pol

[16]http://www.guardian.co.uk/music/musicblog/2009/dec/10/future-fan-funding

SUMMARY

There is a mood afoot in the industry for artists, composers and independent businesses to find alternative funding structures. I applaud these new routes to market that are helping creative talent, especially when the major labels are cutting back, banks are stifling business and piracy is rife. Somehow creativity is finding new funding and entrepreneurial schemes are bubbling up.

SEYMOUR STEIN

Seymour Stein is someone I will never forget and whose kindness always amazes me. Seymour is a great networker and always so generous with introductions; from our earliest collaborations back in 1989 with his writer Jeff Vincent to recent introductions at MIDEM and Musexpo. To Seymour, I'd just like to say 'thank you'.

For those of you who've not had the pleasure of meeting with him, he is a legend and an extraordinary A&R talent. Here's Seymour's background.

Born in Brooklyn, New York, in 1942, Seymour Stein began his music career at *Billboard Magazine* at the age of 13. This experience, coupled with the guidance of Syd Nathan (founder of King Records), taught Stein the workings of the music industry.

By 1966, Stein had set up Sire Records with producer and songwriter Richard Gottehrer. Initially focusing on blues and progressive rock, Sire Records saw its first hit single in 1973 with the band Hocus Pocus. The label's big break came in 1975, when Stein signed both The Ramones and Talking Heads, spearheading the punk and new wave scene.

Since then, Seymour Stein's reputation has continued to grow, and he has signed acts such as Madonna, Lou Read, The Pretenders, The Cure and The Smiths, to name but a few.

Today Stein is still working with new acts under the Sire Label as part of the Warner Brothers label portfolio.

"I wish there were less damn genres", he states. "When I was a kid, there were three categories; pop, country and R&B. Now it's just ridiculous. Because, at the end of the day, there are only really two types of music; good and bad. And that's all you need to know".[17] "Every trend in music in the past 50 years, and probably beyond that, has been started and nurtured by independents".[18] Seymour is a man that has always spoken up for independent labels and producers. He is someone you'd see checking out a band in a dingy club. I remember someone asking him why he went to gigs and why he didn't send his staff. I remember very clearly his answer: "Why would I send someone else to do the most important job there is?" Everyone I know has a story about seeing Seymour at a gig – someone out of the norm.

With the newly formed Sire Records Group (SRG), Stein hoped to build a bridge between two distinct industry powerhouses by cutting distribution pacts with various indies and utilizing distribution from both WEA and Warner Music Group's independent Alternative Distribution Alliance.

Seymour Stein took part in the first music industry panel 'In the City' in New York in 2007 discussing a question fed by Tom Silverman Chief Executive of Tommy Boy records, who asked, "Why haven't we got bold innovative new artists?", to which Seymour replied, "We got lazy".[19]

A short footnote here. New York was a great place back in the 1980s and 1990s for music conferences, which were then called the New Music Seminars. They were brilliant and my network with all the key executives in the USA developed through these conferences.

[17]http://swindlemagazine.com/issueicons/seymour-stein/
[18]http://www.allbusiness.com/retail-trade/miscellaneous-retail-retail-stores-not/4660402-1.html
[19]http://mayareynoldswriter.blogspot.com/2007/06/music-industry-also-debates-its-future.html

Acknowledgements

A big thank you to all those people who have willingly assisted me with their time and have generously given their support to this book. Grateful thanks to my long time and dear friend Alexis Grower for his input, to Simon Napier-Bell, Phil Graham, Anthony Hall, Henri Bellolo, Nigel Elderton, Hawk Wolinski, Freddie Canon, Lawrence Steinberg, Pamela Oland, Javier Lopez, Mike McNally, Ewan Grant and Oliver Sussat. To Peter Rhodes and Javier Lopez: a big thank you for the excellent work that you do and assistance you have given me in supporting new entrepreneurs.

This book is dedicated to my husband, Rod Gammons, who has loved and supported me for over 25 years, and to my children, Helen and Rod, who are forging their own careers, and of whom I am so proud. You remain my biggest inspiration. x

The Art of Music Publishing

By Helen Gammons

Featuring interviews with leading music and media industry executives and legal comment from the industry's leading international lawyers.

This book is designed for those who are:

- studying on a music, media or film course
- working for a music publisher
- interested in setting up their own music publishing company
- working for a 'rights' organization and administering copyrights
- working as a subpublisher or administrator of third party catalogues
- working in all fields of media, and who require content for synchronization or licensing
- wishing to understand the language and structure of the music industry
- working for labels and wanting to understand mechanical payments
- budding artists/writers who want to get a major or an independent publishing deal, and need to know the upsides and downsides of doing either
- wanting to self-publish
- composers wanting to focus on writing but to have a good understanding of music publishing before engaging a manager to represent them, to secure a music publishing contract
- wanting to be more effective within the music industry
- managing a songwriter, writer/artist, writer/producer or lyricist and needing to know about music publishing to get the best for clients
- wanting to know the key areas of negotiation and key commercial points within music publishing contracts
- wishing to become better informed
- wanting to be able to negotiate subpublishing deals and even to buy catalogues, and understand the areas of risk involved
- wishing to put in place an international publishing structure for a catalogue or company
- wishing to be more entrepreneurial and creative in business decisions
- needing to be part of, and engaged with, the music and media industry
- aiming to be successful.

The book is written with the sole purpose of helping you understand music publishing, the industry structure, the development of income streams and the legal terminology that affects its mechanisms and development, infused with entrepreneurial business development. I am not a lawyer but have explained points of detail using examples from contracts. Legal comment is supplied by well-respected industry lawyers. My experience in music publishing and the music industry are my own and examples that I have provided are there to enrich your understanding, and acknowledge the complexity of key areas. Throughout the book I recommend the importance of seeking legal advice for all contractual negotiations. No liability can be accepted by me or the publishers and their licensees for anything done in reliance on the content provided herein – Helen Gammons (MBA).

WHAT THE INDUSTRY SAYS:

"Music publishing? It must be Helen Gammons. She knows all the people, all the gossip, all the stories, all the background. But more importantly, she knows everything there is to know about how music publishing works. And how to make it work for you."

Simon Napier-Bell – Manager, Songwriter, Author (The Yardbirds, Marc Bolan, Wham!
and George Michael, Japan)

"Helen Gammons has been a leading music and publishing executive who has witnessed the profound changes and opportunities that have arisen in the music industry over the past two decades. She has always been known for having a keen eye in developing new musical talent, and this book is an excellent guide through the thicket of today's cross-platform music and music publishing industry – in the UK and internationally."

Ralph Simon – CEO, Mobilium International www.mobilium.com Strategy. Insight. Success.

"In my experience she has an innate understanding of how the music industry works and can unravel the detail of how to get the best from it. Something not many people can achieve, even those in highly placed positions."

Ian Curnow – Established Hit Producer/Writer/Remixer, P+E Music Ltd

"I have known Helen for over 20 years and was inspired by her to pursue a career in the music business which lasted 15 years. I still hear her advice when globally licensing the Rubik's Cube!"

David Hedley Jones MA – Senior Vice President Rubik's Brand, Seven Towns Ltd

"Whenever I have a question about ANYTHING in the music business that I don't already know, the first person I call is Helen Gammons. She always has the right answer!"

David (Hawk) Wolinski – American Grammy Writer/Producer/Keyboardist – 'Ain't Nobody',
Rufus and Chaka Khan, The Bee Gees, Glen Fry, Quincy Jones, Michael Jackson, Chicago,
Beverley Knight

"Helen is a great professional with great experience and global vision of the music industry. She is always a pleasure to work with and has inspired and supported many new talented individuals over the years."

Javier Lopez – REED MIDEM

"Helen was great to work with and licensed out tracks for games use. Top qualities: personable, on time, high integrity."

Greg Turner — Universal Music

"It has been my absolute pleasure to work for and alongside Helen in her role as Head of Business School and Lecturer at the Academy of Contemporary Music, Guildford. Her passion is immeasurable, her commitment unfaltering, and her dedication unquestionable. Her high level entrepreneurial skills coupled with her ability to nurture make her a top-class educator and an inspiration to all."

Oliver Sussat, University Lecturer

"I've known Helen for more years than would be polite to admit, as we continue to cross paths at various international music conferences and conventions. She is an extremely knowledgeable and capable music executive with a special expertise in music publishing."

Lenny Kalikow — Owner, Music Business Reference, Inc.

"Helen has a refreshing 'can do' attitude and I have no hesitation in recommending her and this book."

Bob James — Client Management, Asylum Group

Turning music into a business

The question I am asked more than any other about working in the music industry is, "how do I start?" However, I think they often mean, "how can I short-cut the years you've put into learning your craft, so that I can do it all myself right now?" A-ha, wouldn't that be great! However, to be good at anything in life you have to put the effort in. Knowledge is a great bedfellow, it will enable you to make balanced decisions in your life — so keep reading and talking, and don't ever think that you can stop learning.

The fact that you now have a copy of this book already speaks volumes about who you are and your aspirations. There is no easy fix to your dilemma, so start by taking a brief look at an overview of the industry. It's a bit like a route map to help you understand the direction you need to go in at any one time. The route map fundamentals stay fairly constant, but the economic and strategic pressures bearing down on a business and its strategy for development are constantly being reviewed. Change is occurring so rapidly in the music industry that it certainly keeps learning fun. Of course, if it were as simple as learning to go from A to B, we would have all cracked the industry a long time ago. It is a combination of learning the operational elements that affect each step and then how to create a sustainable business. I will endeavor to share as much with you as possible.

In the Introduction, I referred to Mike Collier. He was a man unfazed by change as he knew the fundamentals: a great song lights up the world and with a good business head you'll find out how to maximize its income stream irrespective of the technological changes in delivery systems or consumer buying behavior.

Figure 1.1 is provided courtesy of the performing right society PRS for Music and provides a really great snapshot of the key companies involved in collecting and creating money for rights owners (those who own the music and the recording). The rights owners are at the core of the industry: songwriters, composers, music publishers, artists and labels. The entire industry revolves around the management of these rights.

Many organizations can assist you in your quest for knowledge, helping you to gain a better understanding of the industry and giving you access to its network. These organizations will be introduced as we go through the book. Here are two that you should consider joining sooner rather than later:

- **The Music Publishers Association** — (MPA) www.mpaonline.org.uk (in the USA: www.mpa.org). The MPA offers training and assistance by way of seminars, website information and networking sessions for music publishers. Don't worry if you're not in the UK; each territory has its equivalent association, so ask your own royalty collection society and they will point you in the right direction.

The Art of Music Publishing. DOI: 10.1016/B978-0-240-52235-7.10001-6

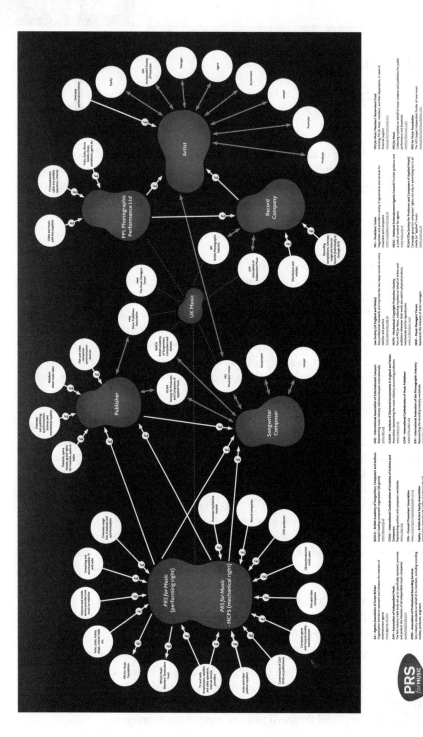

FIGURE 1.1

The music universe in 2010.

- **The Music Managers Forum** — (MMF) www.themmf.net (in the USA: www.mmfus.com). The MMF offers training and assistance in the UK. They cover all aspects that an artist manager will need to know. Importantly, you will be able to network with managers who will be looking to build relationships with music publishers. Lawyers are regular attendees at networking sessions and building relationships with them will broaden your access to the industry in general. Again, every territory will have its equivalent forum.

To understand how best to explore the future, it is essential, especially in music publishing, to understand the operational elements of the industry as they are today. Although the focus of this book is on music publishing, I will elucidate upon the overall structure of the industry to explain some of the many and varied income streams. Outside the UK, currently *Billboard Magazine* gives an essential insight into the USA market and a brief overview of European and other markets. FMQB covers radio and music news in the USA.[1]

Music publishing can be confusing, but when broken down into bite-sized chunks it is a fascinating and very rewarding sector of the music industry. Consequently, there are many individual segments of music publishing that can lead to employment. In addition, you have the opportunity to build your own catalogue and to start up your own business.

WHAT IS MUSIC PUBLISHING?

Berklee Music would say that it's about "exploring the five functions of a music publisher — acquisition, exploitation, administration, collection and protection".[2]

EMI Music would say: "Music publishing is the business of acquiring, protecting, administering and exploiting the 'rights' in musical compositions".[3]

Both of the above are true, but to me it is more than that. Music publishing is not just a job of work, for me it is about hearing a song and seeing how it can live and breathe in many different ways. It is about taking care of your primary assets, which are the song and the songwriter, building a strong relationship with the songwriter and using your best professional endeavors to assist in their career development, whilst creating exciting income streams that allow for a strong company and a happy writer. It's also about integrity!

Because music publishing has, for many companies, become more of a strategic banking operation it has lost much of what music publishers were great at doing, being innovative and passionate. I remember clearly when accountants and lawyers started to take key roles in the creative industries, to the exclusion of 'creative' music publishing employees. Around 1990 we lost a huge number of great music publishers from within major publishing companies. Alan Jacobs at EMI was one

---

[1] www.fmqb.com
[2] www.berkleeshares.com/music.../music_publishing_overview
[3] www.emimusicpub.com

such individual, leaving those who remained with the impossible task of managing burgeoning catalogues through mergers and acquisitions.

It's interesting now, and my good friend Hawk Wolinski (Grammy-nominated writer, composer of 'Ain't Nobody', who in April 2010 received an award for song of the year, covered by Pit Bull using a sample of 'Street Player') agrees with me on this when he says, "It's all about money, greed can kill anything ... but this is a new era for the independent, for creativity for passion and great business".

As a music publisher it is your duty to develop and grow both the skills of the writer (and they could be described as writer/artists, writer/producers, writer/DJs, lyricists) and your income streams, and in doing so build a viable, profitable business for yourself, or those for whom you work. In contract terms this is described simply as a publisher using their 'best endeavors', a phrase of no commitment. I remember the days when Famous Music would commit to a guaranteed number of film synchronizations a year. OK, so they were Paramount Film's publishing division, but Alan Melina was passionate and committed to activating titles.

Creative exploitation (this is a positive phrase) is nothing new; it has always been the ethos of independent publishers. This is not to say that major publishers do not hold true to these values, but their focus is too often on the bottom line first (i.e. their profitability), and creative catalogue development comes second. I am not saying that the bottom line is not supremely important (I am a business owner, and of course it is), but being a creative music publisher will ensure strong and diverse income streams and ultimately a good bottom line will develop. For me, nothing would be worse than if my day were devoid of creativity.

By way of example, the recent purchase of EMI by Guy Hands (Terra Firma[4]) has resulted in a catastrophic number of creative managers leaving or losing their jobs. EMI's artists include, or have included, Joss Stone, Lily Allen, Coldplay, Robbie Williams and The Chemical Brothers.

At the point of purchase EMI's shares were worth 265 pence each. EMI's issued share capital stood at £2.4 billion, and EMI's shares climbed by 3.8% to 263.75 pence in May 2007. The board of EMI agreed to be bought by Terra Firma on 21 May. EMI was one of the world's big four record companies. In May 2007, it said that it had made a £260 million loss in its latest financial year. This has got considerably worse; the record label lost a staggering £1.56 billion in 2009, leaving it in desperate need of a cash injection to prevent Citigroup from taking control of the business.

EMI lost sight of music being about passion, commitment, fun and hard work and, yes, the bottom line. If you remove the passion from a creative business what do you have left? Your assets are the A&R staff, the songwriters, the artists, and the goodwill and commitment of your staff and their talent, and all those

---

[4]Terra Firma Capital Partners is a private equity firm specializing in securitization, based in London, UK. It was formed as a spin-off company from Nomura Group's Principal Finance Group in 1994 by Guy Hands and has invested in excess of £5 billion since its creation, in areas ranging from waste management to ex-Ministry of Defence housing, leisure (pub and hotel ownership and cinemas) and transport, as well as others.

associated with them. If artists walk out or become difficult, or their key marketing or A&R link person is removed overnight it will unsettle them. The artists and A&R staff create the copyrights on which labels and music publishers build their business. Intellectual property (IP) management successfully markets and derives income from such rights for the benefit of the creators and the company.

How much of EMI's current turnover will have come from new signings and new income stream development, and how much from their existing asset base?

When Guy first bought EMI I made some notes as part of a lecture and wrote the following possible scenarios about how the company might progress.

- Should EMI's income start to decline, EMI may start to purchase catalogues (IP) with strong income streams to assist in developing their turnover and market share but at the same time cutting overheads (staff).
- EMI will make it a policy to strengthen the work that EMI Records and EMI Music do together.
- A strategic joint A&R team will be set up, tying EMI Records and EMI Music together to make such decisions.
- They may try to capture and develop talent at an earlier stage (as A&R used to be), costing less upfront and only going to market when a true fan base has been monetized.
- Or they may sign market-ready material: the downside is you have to pay more for it, but there is less risk as it will come with some proof of concept, something tangible, i.e. a fan base. Market-ready material may be considered in the form of established brand names or heritage acts, such as were signed to Sanctuary Music. Heritage acts have an existing fan base, lower risk but lower rewards, and potentially much higher synchronization income.
- If income fails to develop then further funding will be sought.
- EMI will be sold to the first available buyer once the economic recession is over and money is more readily available and a better price offered.
- They may build strategic relationships with the media industry to lead from the front and not to follow, as the industry has done to date.
- They may try to break into new markets, with India being high on the list.
- They will obtain greater income from their existing asset base (most likely from taking advantage of new strategic media relationships, as above).
- The record label may be sold or broken into smaller pieces to keep the publishing company.
- They may try to secure new funding and aggressively purchase music publishing catalogues. Everyone knows that publishing catalogues are bankable deals, whereas labels are a liability!
- Investment in creating new income streams may be achieved by hiring key creative staff in synchronization fields. They may generally be faster and more spontaneous in agreeing deals. Majors have a reputation for being too slow and therefore losing out on too many opportunities.

- Worse possible scenario — they sell the publishing company to clear debts and prop up the label. On paper it might work but there would be little left.

Each possible scenario comes with a balance of risk and reward. This could be described as the upside and the downside to each decision. Scenario planning is all about 'what if', a war-gaming strategy used by the military and an effective business tool to broaden and deepen one's analysis of each strategic option.

Consider when you next need to make a key decision about what your options are. Ask yourself: What is the upside and downside risk for each key decision being contemplated? What are the key forces that could impact on this issue? Rate those forces as to the most likely and least likely. A strategy can then be worked out as to the action or reaction your business should take or make.

It's a shame that the banking community never understood risk analysis, as they would have had to consider how to cope in adverse conditions. It's now part of their training, but it appears that those at the top chose to ignore anything other than a high-reward strategy, irrespective of risk.

Another good example of a failure in scenario planning is BA and the catering strike that grounded all their long-haul flights. How ridiculous that this situation was ever allowed to occur without management having a strategic solution for such a major high-risk event. While trying to achieve cost savings using economy of scale they lost sight of risk and had no back-up plans. The result was their entire long-haul fleet grounded until food and refreshment provisions could be provided at a cost of millions to the airline!

Passion alone cannot ensure that a good decision is made; it has to be passion that drives you, but you need a good decision-making process to ensure financial stability and growth. Guy Hands' defense that it is everyone else's fault other than his own that he bought EMI for too much money doesn't hold water. He can blame the bank and his advisers, but at the end of the day he was in charge and should have known or understood the principles of copyright valuation. If he didn't his advisers should have done. No one would use high multiples when valuing a record label in a declining market being impacted so heavily by music piracy, and for music publishing catalogues, clearly if record sales are declining so too is mechanical income, so this has to be factored in. Nov 4th: It was reported in the press that a New York jury today rejected a fraud case against investment bank Citigroup brought by British financier Guy Hands over his £4.2bn purchase of the EMI music group in 2007. The decision in the US district court in Manhattan could now force Hands's private equity firm, Terra Firma, to relinquish control of EMI to Citigroup, the main creditor bank that provided £2.6bn in loans for the acquistion. We'll look at an element of copyright valuation in Chapter 13.

The development of income streams for the business is just as dependent on your own business skills as it is on having good-quality material to work with. As we can see in the examples below, when you get it right your assets are appreciating — your music catalogue is growing in value.

In 1963, The Beatles originally assigned their publishing rights to a company set up by their manager Brian Epstein and music publisher Dick James, called Northern Songs. The company was floated on the London Stock Exchange in 1965, and in 1969 the Associated Television (ATV) Corporation bought out Dick James and Charles Silver's (Northern Songs' chairman's) share, giving them a controlling stake in the company. In 1985, the ATV Music catalogue, which still included The Beatles' work, was bought by Michael Jackson for a reported £34 million. He later merged the ATV catalogue with Sony Music Publishing, netting a reported £80 million from the move.[5]

More recently, the German media group Bertelsmann agreed to sell its BMG Music Publishing Group to Vivendi's Universal Music in a €1.63 billion (US $2.1 billion, £1.1 billion) deal.[6] Their catalogue includes artists such as Coldplay, Robbie Williams and Justin Timberlake.

As we can see, music publishing catalogues are generally a great asset, whereas the value in the recorded 'master' can only depreciate over time. The internet has ensured this — a topic that featured in my Master's dissertation well before the industry had woken up to the express train heading its way. 'Tightening the net — making the net pay' was a snapshot of the industry as I saw it in 2001, well before the BPI had issued its first loss of income warning that the industry was likely to suffer as a result of internet music piracy. At that time there was no industry strategy and there were no figures. When I met the editor of *Music Week* and offered my report, on reading it he said that he couldn't possibly publish my findings because the level of potential loss was far too high. A few years later the first BPI report was released and was a virtual copy of my own figures, corroborating my findings. A more comprehensive report can be found on the web.[7–9]

No doubt this blinkered approach by the industry came about owing to its fragmented collection of different rights organizations and bodies. I am pleased to say that this has changed somewhat now, with UK Music,[10] whose job is to represent the collective interests of the UK's commercial music industry.

Current members of UK Music include the Association of Independent Music (AIM), the British Academy of Songwriters, Composers, and Authors (BASCA), the British Recorded Music Industry Ltd (BPI), PRS for Music, the Music Managers Forum (MMF), the Music Publishers Association Ltd (MPA), the Musicians' Union (MU) and Phonographic Performance Limited (PPL).

Their core goals are: *to promote awareness and understanding of:*

- the interests of the UK music industry at all levels
- the value of music to society, culture and the economy

---

[5]Information was gathered from: http://news.bbc.co.uk/1/hi/entertainment/music/1323693.stm
[6]http://news.bbc.co.uk/1/hi/business/5319050.stm
[7]http://www.ifpi.org/content/library/The-Impact-of-Illegal-Downloading.pdf
[8]The full article can be seen at www.journals.uchicago.edu/JLE/journal
[9]http://www.ifpi.org/content/library/piracy-report2006.pdf
[10]www.bmr.org and http://www.ukmusic.org/

- intellectual property rights and how they protect and promote creativity
- the opportunities and challenges for music creators in the digital age.

The genie is out of the bottle now: rather than looking at how the industry should have reacted, or how slow the industry was to react, your job is to look at the current landscape and see how you can develop your business or your career in the industry as it is now, and may be in the future. Wow, if only we all had a crystal ball! In that respect this book will never be out of date as change is constant; dealing with it will always require strategic thinking and planning and this will be at the core of this book, working alongside all that is 'music publishing'. There are often clear indicators illuminating possible choices and decisions (paths and opportunities). Indeed, 'scenario planning' is a good-quality strategic tool to have in your business armory. A great selection of business tools will be reviewed later in the book.

THE A&R PROCESS

So how do you start? Why not start with the basics? Take a leaf out of Seymour Stein's book and go out and experience music for yourself. If you do this already (and I'm sure you do) then try to develop your skills. Start listening to music and start doing some primary research.[11] Discover who's writing the songs (this could be the whole band, or it may just be one member). Find a song or a writer that really excites you and ask yourself why. There is a variety of songwriter and performer nights all over London, New York, Los Angeles, some being industry-led nights organized by the American Society of Composers, Authors and Publishers (ASCAP), Broadcast Music Inc. (BMI), Robertson Taylor Music, PRS for Music and others.

Decide what is it that you like. Could others like what you hear? What would you have to do to translate that song or writer into an income stream? Can you build a business around those copyrights and the composer? Do you think the writer has long-term talent? Does the song you like inspire you with ideas as to how it could be used? Perhaps other artists could do a great version of it? Is it suitable for use in a computer game? Which game would it work in? Can you track down the computer gaming company and the person who is engaged to develop the project? How are they sourcing the music? Do you think the songs could work in an American TV soap, and if so, which one? Does it have a great sports connotation? A lyric that could inspire being used in sports may talk about reaching for the top, achieving, pushing, winning …. We've just had the World Cup. What other sporting events are coming up, and could television welcome your music?

There are many ways to generate income from a great song or a great piece of music.

---

[11]Primary Research is information that did not exist before the research was carried out. It is data specifically relating to your product or service, designed to answer questions that have not been answered before.

INCOME STREAMS

So, what are the key areas of income stream for music publishers and songwriters? Simply speaking, this breaks down into performing rights and mechanical income.

- **Performing rights** income is generated when music is broadcast or performed in public. This is a much broader arena nowadays, including internet, streaming, webcasting and podcasts, and delivery formats will keep evolving. It includes performances and broadcasts on radio and television and in theater, live concerts and tours, and venues licensed to play music (bars, hotels, clubs, retail stores, etc.). For example, every time a song is played on the radio it generates money. Currently, on BBC Radio 1 in the UK it's about £60 for a three-minute play. On smaller stations it's a lot less. A performance of a song on national television could generate, as a guide, £300 (less on local stations).
- **Mechanical income** is money generated from sales of records, in a physical or digital format. A percentage on the sale of each record goes to the copyright owners of the song. Mechanical income is also derived from the sale of music in an evolving variety of formats, including videos, ringtones, DVDs and musical toys.
- **Synchronization** — Both the copyright owners of the song (music publishers) and the owners of the master recording (record label) negotiate a fee to allow a specific use. This is music placed in a timed relationship to a visual image. The most likely outlets at present are film, television and film advertising commercials, and computer games. The sort of sums that get paid vary greatly depending on whether the song is known or unknown, is featured or in the background, and how long the music is played for (this might be just seconds or minutes). Fees could vary from £250 to £100,000 or more.
- **Print rights** are due when lyrics are published in a magazine, when sheet music is sold or when a book containing a selection of songs (portfolio) is sold. Books are now both physical and digital and so too are magazines.

SUMMARY

Start by:

- developing A&R skills
- reading about the industry — subscribe to key industry magazines and get RSS feeds from key technology sites
- joining key industry organizations
- starting to build your network
- developing a list or database of customers, different media organizations that need music, artist managers and record label personnel
- finding a song or a writer to focus your attentions on — it could be your own material.

Have you found a writer or writer/artist who you feel consumers will like? How have you proved this? Are gigs packed, and is there a strong reaction online to the material? What feedback are you getting from the press? Is this someone you can work with?

You may be considering representing your own catalogue and becoming your own publisher (more established writers tend to do this). Decide whether you have time to do this or whether you need a publisher to be creative or just to administer you.

Throughout the book you'll read various life stories of business entrepreneurs and passionate music people who made a difference. Here's a great one for you.

HENRI BELOLO – FRENCH ENTREPRENEUR, WRITER/ PRODUCER/LABEL/PUBLISHER

I've known Henri since the 1980s and we've done a fair amount of business together over the years. I traveled to Paris in 2009 to interview him, the outcome of which will be available online soon.

In brief, Henri Belolo was born in 1936 in Morocco. He studied at University in Paris, France, where, after a chance meeting, he struck up a friendship with Eddie Barclay, president of Barclay Records. He was offered the opportunity to represent the entire Barclay Records' and Atlantic Records' catalogues in Morocco, and ever the entrepreneur, Henri moved back to Morocco and set up his own independent label. Henri's ability to build a market didn't go unnoticed, and amidst political and economic unrest in Morocco, he was invited to move back to France, where he was hired by Polydor's president.

In the few years that followed, he worked in all areas of the business, and within five years he took over as Polydor's youngest ever president. Under Henri's control they had the best years ever. This success was to be the trigger that Henri needed to set up in business on his own. Making vast sums of money for other people was now to be a thing of the past, as the real business entrepreneur had been born. Using $10,000 (the total amount of his savings) he formed Carabine Music, which was to specialize in jazz and classical genres, focusing on the back catalogue market. This brought early success and provided the financial stepping stone for his next adventure.

In 1973 Henri flew to America.[12] During his time in the USA he became friends with the owner of the now infamous Sigma Sound Studios in Philadelphia. It was here that he was introduced to the "driving, full-bodied, orchestral approach of funky rhythms and swirling strings"[13] known as the 'Philly Sound', which was made famous by songwriters Gamble and Huff, and the Mother Father Sister Brother orchestra.

Before leaving to go back to Paris, Henri set up the record company Can't Stop Productions and promised himself that he would return to the USA.

The catalyst for Henri's return came at a fateful meeting in Paris, where he met Jacques Morali. Jacques was looking for a way to break out of France and move to America. Henri saw in Jacques something that excited him, and having told Jacques to come up with a suitable idea to take to the USA, he returned with the concept of turning Brazil by Carmen into a disco song. Henri immediately jumped at the idea, and on returning to Sigma Sounds in Philadelphia the record was produced. With it selling over one million copies, Jacques and Henri's producer/songwriting partnership was founded, and they looked to see what else they could create.

It was while having a few drinks in a bar in New York that their next idea was conceived … The Village People. They wanted to create a group of characters that represented "the mix … of the American man",[14] and having placed an advert in the newspaper, the band was formed. Their first album became an underground success and sold 100,000 copies. Realizing this success, Henri and Jacques signed it over to Neil Bogart at Casablanca Records, who had already had successes with Donna Summer and Kiss. Jacques and Henri's songwriting success continued to grow, transforming the club and disco scene in the 1970s with hits such as 'Macho Man', 'In The Navy' and the all time classic, 'YMCA'.

It was another of Jacques and Henri's songs, 'Go West', which, although initially written for The Village People, was perhaps made most successful by the Pet Shop Boys, reaching No. 1 in Germany and No. 2 across most of Europe. And, 1979 proved to be the year in which disco peaked. Jacques and Henri had taken The Village People on tour, selling out venues such as Madison Square Gardens in New York and The Felt Forum in Los Angeles, while a film charting the rise of the group had also been made. Unfortunately, the film suffered in America with the death of disco but went on to become a great success in Australia.

To this day, The Village People are still touring with in excess of 150 dates a year to the tune of $2 million. The merchandising and brand name rights are both owned by Henri's companies.

Henri has since moved back to Paris and has concentrated on Scorpio Music, having hits with 2 Unlimited, Bass Bumpers, Rosie Gaines, Joan Jett, Haddaway, Fun Factory, Alice DeeJay, Eiffel 65, Phats'n'Small and Spiller. He is still actively involved in the music industry and continues to grow Scorpio Music with his sons following in his footsteps. His office, studio and home are filled with his achievements.

[12]http://www.disco-disco.com/tributes/henri.shtml
[13]http://www.70disco.com/mfsb.htm
[14]http://www.disco-disco.com/tributes/henri.shtml

DAVID PAKMAN – FROM MUSICIAN AND COMPOSER TO IT FUTURIST AND MULTIMILLIONAIRE

David is an inspiring man (and great drummer, having worked with my husband and lead singer Tony Fenell) with an appetite for cutting-edge technology and market vision which led to the growth of a hugely successful business. He now works in the exciting world of investment funding with technology media companies. We last spoke at Musexpo in Los Angeles, in 2007. Here's his story.

David Pakman, CEO eMusic[15]

David joined eMusic as Chief Operating Officer in 2004 and was elevated to CEO in August 2005. "I'm proud of what we've accomplished over the past five years", said Pakman. "eMusic has one of the best operating teams in digital entertainment, has grown into a sophisticated direct marketing enterprise, and is a company with a proven business model. Our subscriber base now tops 400,000 – more than six times where we started, and we've increased revenue tenfold. Our success has outlasted many other players in the space, and we're now the only remaining stand-alone digital entertainment retailer". "David has been instrumental in transforming eMusic from a niche music business to a global direct marketing platform capable of delivering any form of downloadable digital media", said Danny Stein, Chairman of eMusic and President of JDS Capital Management. "While we're proud of David's success and the great team that survives him, we look forward to working with a new CEO who will take the company from $70 million of revenue to several hundred million of revenue".

About eMusic

eMusic (http://www.emusic.com) is a specialty digital entertainment retailer that has been at the forefront of offering MP3 downloads and customer-friendly prices since its inception in 1998. The company is focused on serving customers aged 25 and older by offering independent music and audio books in a universally compatible format at a great value. It was the world's largest retailer of independent music and the world's second largest digital music service after iTunes, with more than four million tracks from 40,000 of the world's leading independent labels and thousands of titles from top audio book publishers. To super-serve its more than 400,000 customers, eMusic provided award-winning editorial content, a vibrant online community and unrivaled music discovery tools. eMusic's subscription-based service offered free music downloads or one free audio book at sign-up, giving consumers an inexpensive, low-risk way to explore great new music and books they wouldn't find otherwise. The company's music download promotions help world-class brands reward and engage with their customers. Based in New York with an office in London, eMusic is available in the USA, Canada and 26 European Union (EU) nations. eMusic.com Inc. is wholly owned by Dimensional Associates Inc., the private equity arm of JDS Capital Management Inc.

New York, 29 September, 2008 – eMusic, the world's largest retailer of independent music and the world's second largest digital music service after iTunes, today announced that President and Chief Executive Officer David Pakman will depart the company. Pakman will become a partner at a premier venture capital firm.

[15]http://www.emusic.com

Beginning to feel inspired? Don't go away, your future's out there.

A business has to be involving, it has to be fun, and it has to exercise your creative instincts.[16]

Richard Branson

---

[16]Branson, R. *Losing My Virginity.*

Understanding copyright

WHAT IS COPYRIGHT?

Let us begin by looking at the technical definitions and structure of copyright, before seeing how this may affect you.

Copyright is a protection that covers published and unpublished works. It exists at the point of creation, arising automatically. The copyright work, however, must exist in a material form, for example, that of a recording or sheet music.

Copyright subsists in the following works:

- original music works
- original artistic works
- original literary works
- original dramatic works
- films, sound recordings, broadcasts, cable programs, typographical arrangements of published editions.

The Copyrights Designs and Patents Act (CDPA) 1988 gives certain economic and moral rights to such works.

HOW LONG DOES COPYRIGHT LAST?

Copyrights exist in the UK and Europe for 70 years after the death of the composer. The creators of the song and their heirs can enjoy the income that the song may create for 70 years after the death of the last surviving writer (if it was written by more than one person). So the song you have is protected by the existence of copyright at the point it is created and exists in material form. In the music industry and often in the press the words song and copyright are often interchangeable. Publishers may refer to their catalogues as consisting of a number of songs, or owning a number of copyrights.

Surely once copyright is established, why should it cease to exist after a number of years? Why on earth after a period of time should the composer of the work no longer be acknowledged or receive an income from his or her work? It is quite ridiculous, and composers and the industry at large must endeavor to get this changed. The anti-copyright extension lobby argues that the increased copyright term favors only big corporations. Not so; yes, of course, existing publishers benefit if they control works for 'life of copyright', but these days fewer new writers are

The Art of Music Publishing. DOI: 10.1016/B978-0-240-52235-7.10002-8

13

signed on this basis. The beneficiaries are the original copyright owners (the composers) or their heirs (family). The only right answer is to make sure that when a song is created composers can lay title and claim to their work forever, and do with it as they wish. Remember this is the writer's income, this is how he or she lives and feeds their family and builds a pension and family worth. Why should this not be handed down from generation to generation? Why is there such a lack of respect given to copyright creators?

On the other side of the copyright coin, regarding master recordings, artists and record labels can see their work become public domain 50 years after the release of the recording. We have already seen recordings by Elvis Presley fall out of copyright in the UK (not in the USA — American copyright is summarized in Figure 2.1).

MASTERS AND RE-RECORDING RESTRICTIONS

Artists generally have the right to re-record their hits once their re-recording restriction has elapsed (usually somewhere between 10 and 20 years after the first release of the work, or a number of years after the conclusion of their contracted term). Many artists coming back into the marketplace are able to take full advantage of this taking control of their recordings (obviously they have to fund the new recordings themselves). If the artist is also the songwriter it puts them in control of both sides of copyright income: money derived for the copyright owners of the master and money derived for the copyright owners of the song. Artists are no longer reliant on record labels for financial support. Artists are now doing direct deals with interactive entertainment companies, telecommunications companies, equity investors, live agents and promoters. There are now many alternative-funding models to bring product into the marketplace. So while the record industry is shrinking and new competitors are entering this space (e.g. iTunes, O2, Nokia, Live Nation, EA Games) who can release and distribute recordings, music publishers are well placed to handle all new areas of income stream development.

Just staying with master recordings for a second, you might also ponder on this point of view from Mick Hucknall (formerly the lead singer in Simply Red, but now a solo artist): if an artist eventually pays back the label who has invested in him for the development of the recordings to the point where that debt is repaid, surely then the ownership of that property, i.e. the recordings, should vest in the artist? I agree it's a fair position to take, but I wonder how many artists reach that point. If they do, surely their royalty should increase dramatically or they should see a reversion of those masters at some point. However, labels like having their cake and eating it.[1]

Let's just go back one step and get the basics clear. Copyright exists in the song and also in the master recording. You'll recall from the Music Universe image in Chapter 1 (Figure 1.1) that the artist and the record company control the master rights. The writer and the music publisher control the song or music rights.

---

[1]http://www.simplyred.com/

Copyright Term and the Public Domain in the United States
1 January 2010[1]

Never Published, Never Registered Works[2]		
Type of Work	***Copyright Term***	***What was in the public domain in the U.S. as of 1 January 2010***[3]
Unpublished works	Life of the author + 70 years	Works from authors who died before 1940
Unpublished anonymous and pseudonymous works, and works made for hire (corporate authorship)	120 years from date of creation	Works created before 1890
Unpublished works when the death date of the author is not known[4]	120 years from date of creation[5]	Works created before 1890[5]
Works Registered or First Published in the U.S.		
Date of Publication[6]	***Conditions***[7]	***Copyright Term***[3]
Before 1923	None	None. In the public domain due to copyright expiration
1923 through 1977	Published without a copyright notice	None. In the public domain due to failure to comply with required formalities
1978 to 1 March 1989	Published without notice, and without subsequent registration within 5 years	None. In the public domain due to failure to comply with required formalities
1978 to 1 March 1989	Published without notice, but with subsequent registration within 5 years	70 years after the death of author. If a work of corporate authorship, 95 years from publication or 120 years from creation, whichever expires first
1923 through 1963	Published with notice but copyright was not renewed[8]	None. In the public domain due to copyright expiration
1923 through 1963	Published with notice and the copyright was renewed[8]	95 years after publication date

FIGURE 2.1

Summary of American copyright.

(© 2010 Peter B. Hirtle. Used with kind permission.)

1964 through 1977	Published with notice	95 years after publication date
1978 to 1 March 1989	Created after 1977 and published with notice	70 years after the death of author. If a work of corporate authorship, 95 years from publication or 120 years from creation, whichever expires first
1978 to 1 March 1989	Created before 1978 and first published with notice in the specified period	The greater of the term specified in the previous entry or 31 December 2047
From 1 March 1989 through 2002	Created after 1977	70 years after the death of author. If a work of corporate authorship, 95 years from publication or 120 years from creation, whichever expires first
From 1 March 1989 through 2002	Created before 1978 and first published in this period	The greater of the term specified in the previous entry or 31 December 2047
After 2002	None	70 years after the death of author. If a work of corporate authorship, 95 years from publication or 120 years from creation, whichever expires first
Anytime	Works prepared by an officer or employee of the United States Government as part of that person's official duties. [21]	None. In the public domain in the United States (17 U.S.C. § 105)
Works First Published Outside the U.S. by Foreign Nationals or U.S. Citizens Living Abroad [9]		
Date of Publication	*Conditions*	*Copyright Term in the United States*
Before 1923	None	In the public domain (But see first special case below)
Works Published Abroad Before 1978 [10]		
1923 through 1977	Published without compliance with US formalities, and in the public domain in its source country as of 1 January 1996 (but see special cases) [20]	In the public domain
1923 through 1977	Published in compliance with all US formalities (i.e., notice, renewal) [11]	95 years after publication date
1923 through 1977	Solely published abroad, without compliance with US formalities or republication in the US, and not in the public domain in its home country as of 1 January 1996 (but see special cases)	95 years after publication date
1923 through 1977	Published in the US less than 30 days after publication abroad	Use the US publication chart to determine duration

FIGURE 2.1—*Continued.*

1923 through 1977	Published in the US more than 30 days after publication abroad, without compliance with US formalities, and not in the public domain in its home country as of 1 January 1996(but see special cases)	95 years after publication date
Works Published Abroad After 1 January 1978		
After 1 January 1978	Published without copyright notice, and in the public domain in its source country as of 1 January 1996 (but see special cases)[20]	In the public domain
After 1 January 1978	Published either with or without copyright notice, and not in the public domain in its home country as of 1 January 1996(but see special cases)	70 years after death of author, or if work of corporate authorship, 95 years from publication
Special Cases		
1 July 1909 through 1978	In Alaska, Arizona, California, Hawaii, Idaho, Montana, Nevada, Oregon, Washington, Guam, and the Northern Mariana Islands ONLY. Published in a language other than English, and without subsequent republication with a copyright notice [12]	Treat as an unpublished work until such date as first US-compliant publication occurred
Before 19 Aug. 1954	Published by a Laotian in Laos[18]	In the public domain
Between 18 Aug. 1954 and 3 Dec. 1975	Published by a Laotian in Laos[18]	Use the US publication chart to determine duration
Prior to 27 May 1973	Published by a national of Turkmenistan or Uzbekistan in either country[19]	In the public domain
After 26 May 1973	Published by a national of Turkmenistan or Uzbekistan in either country[19]	May be protected under the UCC
Anytime	Created by a resident of Afghanistan, Eritrea, Ethiopia, Iran, Iraq, or San Marino, and published in one of these countries[13]	Not protected by US copyright law until they become party to bilateral or international copyright agreements
Anytime	Works whose copyright was once owned or administered by the Alien Property Custodian, and whose copyright, if restored, would as of January 1, 1996, be owned by a government[14]	Not protected by US copyright law
Anytime	If published in one of the following countries, the 1 January 1996 date given above is replaced by the date of the country's membership in the Berne Convention or the World Trade Organization, whichever is earlier: Andorra, Angola, Armenia, Bhutan, Cambodia, Comoros, Jordan, Korea, Democratic People's Republic, Micronesia, Montenegro, Nepal, Oman, Papua New Guinea, Qatar, Samoa, Saudi Arabia, Solomon Islands, Sudan, Syria, Tajikistan, Tonga, United Arab Emirates, Uzbekistan, Vietnam, Yemen	

FIGURE 2.1—*Continued.*

Sound Recordings		
(Note: The following information applies only to the sound recording itself, and not to any copyrights in underlying compositions or texts.)		
Date of Fixation/Publication	*Conditions*	*What was in the public domain in the U.S. as of 1 January 2010* [3]
Unpublished Sound Recordings, Domestic and Foreign		
Prior to 15 Feb. 1972	Indeterminate	Subject to state common law protection. Enters the public domain on 15 Feb. 2067
After 15 Feb. 1972	Life of the author + 70 years. For unpublished anonymous and pseudonymous works and works made for hire (corporate authorship), 120 years from the date of fixation	Nothing. The soonest anything enters the public domain is 15 Feb. 2067
Sound Recordings Published in the United States		
Date of Fixation/Publication	*Conditions*	*What was in the public domain in the U.S. as of 1 January 2010* [3]
Fixed prior to 15 Feb. 1972	None	Subject to state statutory and/or common law protection. Fully enters the public domain on 15 Feb. 2067
15 Feb 1972 to 1978	Published without notice (i.e, , year of publication, and name of copyright owner) [15]	In the public domain
15 Feb. 1972 to 1978	Published with notice	95 years from publication. 2068 at the earliest
1978 to 1 March 1989	Published without notice, and without subsequent registration	In the public domain
1978 to 1 March 1989	Published with notice	70 years after death of author, or if work of corporate authorship, the shorter of 95 years from publication, or 120 years from creation. 2049 at the earliest
After 1 March 1989	None	70 years after death of author, or if work of corporate authorship, the shorter of 95 years from publication, or 120 years from creation. 2049 at the earliest
Sound Recordings Published Outside the United States		
Prior to 1923	None	Subject to state statutory and/or common law protection. Fully enters the public domain on 15 Feb. 2067

FIGURE 2.1—*Continued.*

1923 to 1 March 1989	In the public domain in its home country as of 1 Jan. 1996 or there was US publication within 30 days of the foreign publication (but see special cases)	Subject to state common law protection. Enters the public domain on 15 Feb. 2067
1923 to 15 Feb. 1972	Not in the public domain in its home country as of 1 Jan. 1996. At least one author of the work was not a US citizen or was living abroad, and there was no US publication within 30 days of the foreign publication (but see special cases)	Enters public domain on 15 Feb. 2067
15 Feb. 1972 to 1978	Not in the public domain in its home country as of 1 Jan. 1996. At least one author of the work was not a US citizen or was living abroad, and there was no US publication within 30 days of the foreign publication (but see special cases)	95 years from date of publication. 2068 at the earliest
1978 to 1 March 1989	Not in the public domain in its home country as of 1 Jan. 1996. At least one author of the work was not a US citizen or was living abroad, and there was no US publication within 30 days of the foreign publication (but see special cases)	70 years after death of author, or if work of corporate authorship, the shorter of 95 years from publication, or 120 years from creation
After 1 March 1989	None	70 years after death of author, or if work of corporate authorship, the shorter of 95 years from publication, or 120 years from creation
Special Cases		
Fixed at any time	Created by a resident of Afghanistan, Eritrea, Ethiopia, Iran, Iraq, or San Marino, and published in one of these countries[13]	Not protected by US copyright law because they are not party to international copyright agreements
Fixed prior to 1996	Works whose copyright was once owned or administered by the Alien Property Custodian, and whose copyright, if restored, would as of 1 January 1996 be owned by a government[14]	Not protected by US copyright law
Fixed at any time	If fixed or solely published in one of the following countries, the 1 January 1996 date given above is replaced by the date of the country's membership in the Berne Convention or the World Trade Organization, whichever is earlier: Andorra, Angola, Armenia, Bhutan, Cambodia, Comoros, Jordan, Korea, Democratic People's Republic, Micronesia, Montenegro, Nepal, Oman, Papua New Guinea, Qatar, Samoa, Saudi Arabia, Solomon Islands, Sudan, Syria, Tajikistan, Tonga, United Arab Emirates, Uzbekistan, Vietnam, Yemen	

FIGURE 2.1—*Continued.*

Architectural Works [16]		
(Note: Architectural plans and drawings may also be protected as textual/graphics works)		
Date of Design	*Date of Construction*	*Copyright Status*
Prior to 1 Dec. 1990	Not constructed by 31 Dec. 2002	Protected only as plans or drawings
Prior to 1 Dec. 1990	Constructed by 1 Dec. 1990	Protected only as plans or drawings
Prior to 1 Dec. 1990	Constructed between 30 Nov. 1990 and 31 Dec. 2002	Building is protected for 70 years after death of author, or if work of corporate authorship, the shorter of 95 years from publication, or 120 years from creation [17]
From 1 Dec. 1990	Immaterial	Building is protected for 70 years after death of author, or if work of corporate authorship, the shorter of 95 years from publication, or 120 years from creation [17]

FIGURE 2.1—*Continued.*

This is very important as there are separate organizations that control or represent either side of the industry. This is the one area that so many individuals get wrong. When music is cleared for use for a film synchronization, or a television synchronization, or another artist samples someone else's record, there is more than one set of permissions to get cleared and therefore there are twice as many contracts to enter into. Negotiations, permissions and contracts have to be entered into with all rights owners of the entire copyright (song and recording) (Figure 2.2).

It follows that if you write and produce your own material (and many composers and artists now do) then you potentially own the copyright of the song and the master. If you have co-written with someone else, or you have worked together to produce and pay for the recording (or agree for this to be collectively owned) then the copyright is shared between the participants.

To clarify — if you co-wrote the song with someone then each composer would share in the copyright of the song. The writers would have to reach an agreement on how the song is split, and what percentage of the song they would each control. If you equally shared the writing then this would be 50% to each writer. Each writer can then decide independently how he or she wishes to have his or her share of the song represented. You can each select to be published by a different music publisher who will collect your share of the song for you. For example, 'Sister Sister', recorded by Beverley Knight, was written by Rod Gammons (G2 Music/Peer), David Hawk Wolinski (Rondor) and Beverley Knight (EMI). Each writer owned one-third of the song.

To verify this, go and read the label copy on your album and see for yourself who the writers are and who publishes each writer. On a lot of hip-hop titles where sampling is prevalent this can involve a very long list of publishers as each sample must recognize (and pay) each owner of the copyright involved in that track. If the

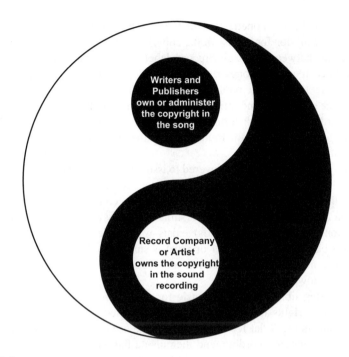

FIGURE 2.2

Copyright in music — two parts working together. Copyright exists in the composition (song or music) and copyright also exists in the sound recording.

samples are cleared in advance of being used there is every possibility of an amicable agreement being reached. However, if samples are not cleared then, in general, only the original copyright owners end up receiving all the money from the exploitation of the song.

OWNERSHIP OF THE MASTER RECORDING

For example, if you have created and paid for the recording as a band then you might agree, among all of you, that you all collectively own the recording and the copyright in that recording and therefore any money you generate from it will be shared equally. If an artist signs to a record label, the label will own the copyright in the master recordings as they generally front the money to pay for them.

U2 agreed from day one that irrespective of who writes which bit of the song they would share equally in the income derived from all exploitation of the songs. This has created a harmonious band relationship. This is not always the case, however, and each band or artist must decide for themselves what is best.

So you can see that you can have different owners of the song from those of the recording. It is often this point that really confuses people, but really it's quite straightforward. In the monetization of these rights all copyright owners have to supply their permission for the use of the copyright. This is usually done through the contract either between artist and record label, or between composer and music publisher. In each case the songwriter or the artist gives permission under the terms of their contract to the music publisher and/or the label, so that in simple terms they can act for the songwriter and artist in the exploitation of the copyright in the music and the recording. We'll look at this in more detail later, in the contract section. You can now begin to understand the foundations of the industry and the monetization of these rights.

The copyright owners are empowered to issue licences for the use of the songs or masters to whoever wishes to use them. Such areas might include advertising agencies, film companies, television companies, computer games companies, distribution companies and others. Vast sums of money are created and an industry worth billions of dollars (or pounds) exists worldwide. How the money that is created gets back to the songwriter or artist is subject to their contracts. For song-writers, music publishers and collection societies are the established route through which money flows.

Those writing library music or music for television and adverts will typically own all their own rights and those rights will be licensed or acquired for specific use by media companies. A flat fee is payable for the use (synchronization licence) or purchase of the master recording, and then also a flat fee for the use of the music/song copyright. In addition, performance income will follow to the song/music composers. This money is collected by PRS for Music (see below) and paid to its members. Its members are music publishers and writers. Irrespective of a songwriter being represented by a music publisher, the songwriter will always get 50% of any performing income directly if they are a member of a performing right society. Their publisher will collect the rest and account to the songwriter on what is due under the terms of their contract. Should a songwriter have a 70/30 agreement with the publisher then they will receive 50% of all performing income directly and the other 20% will be directly accounted to the songwriter from the publisher. This allows the publisher to use the 20% to recoup any monies it may have advanced to the song-writer. The music publisher will receive 100% of all mechanical income directly from the mechanical rights society; in the UK this is the Mechanical Copyright Protection Society (MCPS). The publisher will use the writer's 70% to recoup any advances it has paid to the writer. Since 2009, the Performing Right Society and the Mechanical Copyright Protection Society (PRS and MCPS) in the UK have been jointly known as PRS for Music. Books published before 2009 will refer to them separately.

PPL[2] income can also be collected for the performance of the recording. This is payable to the owners of the master recording who, in turn, pay a proportion to the artists. When the recording is played on the radio, performance income is payable by

---

[2]http://www.ppluk.com — PPL licences sound recordings and music videos for use in broadcast, public performance and new media.

that station to the performing rights collecting society which, in turn, pays the copyright owners (writers/publishers). Performance income is also payable for use of the recording and this is payable by the radio station to the record label via PPL, which collects and administers this income.

In the USA a writer and publisher can choose to have this performance income collected by BMI, ASCAP or SESAC. I have worked with two of these societies, but they all do an excellent job, so it's really down to the people, the ethos and the culture of the organization, and all three societies are quite different in these respects.

Now just stop and think about the above for one minute …

The entire industry revolves around the management of the rights of songs and recordings. If you own both then that's a very strong hand to be holding, is it not? Therefore, taking this thought process on one step, artists and songwriters in today's climate are therefore actually in the driving seat in many respects. However, the ownership of such rights is not enough; you need to have a good helping of business acumen to be able to exploit these rights commercially. Irrespective of the genre of music — classical, rock, indie, punk, R&B, pop, folk, country, whatever style of music you like or create — the road map through the industry and the business acumen needed are the same.

Consider all the areas in which you could do business: film, television, computer games, library music, advertising commercials, record sales, live, in fact all forms of media entertainment. The commercial benefit is that you own both sides of the income stream, i.e. money generated from the master and the song. As new areas of media exploitation or devices are developed, the areas via which royalties, flat fees, buy-outs, subscription fees, performance fees, mechanical fees, streaming rates and more can be received also increase.

LYRIC WRITING — A LOST ART

On the art of writing a great lyric, Pamela Oland, a new friend and fantastic American lyricist who has written for Whitney Houston, Frank Sinatra, Earth Wind and Fire, Aretha Franklin and The Jacksons, to name but a few, has kindly provided the following:[3]

Most people seem to hear melody first. But after they get over that, they want to know what the song says. And if the lyrics don't sell it, then there's only half a song there. Lyrics are a craft of their own. They are a separate area of expertise from music. Being a great musician, singer, player, arranger, producer, doesn't guarantee language skills. Nor do language skills guarantee musical talents or the ability to hear melody. I always teach writers to think of what the listener will hear, not just say what they want to say. Advice? Build

[3]www.pamoland.com

a story, make sure it grows and the listener longs to hear more. I believe lyric writing is very intuitive. I also love to say, "Never let the truth get in the way of a great idea!"[4]

Pamela goes on to say that in America www.songsalive.com is a wonderful organization, helping writers to band together to meet and critique each others' work. Another key organization for writers is the Nashville Songwriters Association International, which is headed up by Bart Herbison. This is a very active organization in lobbying Congress for copyright reform and songwriter/publisher rights.

With your feet firmly back on the ground again, consider also that the real art to making money is considering what your own strengths are.

Typically, sites such as MySpace are really about self-marketing and not a way to make money (at this time). It is interesting to note that in 2010 MySpace and PRS for Music have agreed a strategy of co-interests in providing a safe and creative environment for musicians and bands to promote and share all that they do, and at the same time for PRS for Music to promote the value of copyright and engage in discussing with new writers about membership to PRS for Music.[5] As a writer, ask yourself: do I have the contacts to network myself and my music to the right people? Do I have the time to devote to building those contacts? This is a strategic decision to make about how best to develop your future: build your own publishing company and direct its future, or be published by a music publisher and draw on their strengths? Would it be better for you to focus on writing music rather than building your own business and publishing company?

Essentially these are the first questions to ask and the answers will help to clarify your next step. To help you make that decision and develop accordingly you need to know more about the structure of contracts and how they work, and how you can make them work to your advantage, no matter who you are, the publisher or the composer.

Just one final comment before we move on. Either of the above strategies is great! Neither of them is set in stone. You can start your business or career in one guise and develop later in another way. Many writers start by being published and when their rights revert to them (i.e. everything is back in their control) they form their own music publishing company. There is flexibility, provided that your contracts are always structured well. It is vital to see a good-quality music lawyer.

A songwriter typically enters into a contract with a publisher of good standing and industry experience to represent their songs for a specific period. This is usually structured under an Exclusive Writer Agreement. This is a very serious contract and can tie the writer to the publisher for a fairly long period. Often the composer is looking to receive an advance and some 'hand-holding' in terms of creative development and networking.

---

[4]Pamela Phillips Oland, songwriter and author of *The Art of Writing Great Lyrics* and *The Art of Writing Love Songs* (New York: Allworth, 2001, 2004).
[5]www.prsformusic.com

However, if you wish to just have one or a specific number of songs represented this can be done under a Specific Agreement, often referred to as a Single Song Assignment. In this type of contract the writer is not obligated to the publisher to work exclusively for them. It is just a specific contract relating to titles the publisher will represent for a set period (laid down by the terms of the contract, all of which are negotiable). You should always be represented by a good-quality music industry lawyer throughout any negotiation.

The big change in today's market sees many writers choosing to self-publish in their own country (set up their own publishing company) and then perhaps choosing to formulate a structure outside their home territory by working with a variety of other publishers.

How is this done, I hear you ask? Well, this type of arrangement would provide for the appointment of subpublishers or administrators in other countries. You can elect to choose a different company in each territory, or elect to have Europe taken care of in one go. It is down to you to decide what structure works for you, what you want to get out of the relationship and what you are hoping to achieve. At this time, it can be helpful to write down your aims and objectives, or your company's aims and objectives, and perform a SWOT (strengths, weaknesses, opportunities, threats) analysis to clarify and define the answers to the above and allow you to formulate a strategy moving forwards (see Chapter 14).

For instance, you may decide that you want to build a strong network of people who can help you in all the key territories. By doing deals territory by territory this will allow you to select the strongest companies to work with in any given country: France, Germany, Spain and Italy, for instance. The type of relationship that you want will help to clarify the type of contract that is best for you. Do you need local, creative A&R, local writer and artist introductions, potential licensee introductions? Then you'll need a good, hands-on subpublisher who is prepared to do this for you (and many are). In addition, within the contracts there is scope to shape the relationship further by negotiating the detail.

SWOT ANALYSIS

Strengths

There are many upsides to this arrangement:

- It allows you to choose the best company in each territory.
- It lets you build a relationship directly with each company.
- You can get to know the culture and people's needs and likes in each territory through feedback from the local company.
- You can build a network of writers and creatives through these relationships in each territory.
- There is no cross-recoupment from one territory deal to the next.
- It gives you the ability to encourage each territory to assist you.

Weaknesses

Potential downsides include:

- administering this number of relationships and contracts
- increased legal costs
- co-ordination across a number of territories.

As an example, take pan-European licensing. Nokia wants to use your song across Europe for a campaign. You have eight different subpublishers covering Europe, who technically would have to get agreement and pay eight publishers, unless your contract left synchronization clearance to you, save for local use only or similar. The Gowers Report and European Parliament want to see greater ease of use in such situations. So now rights owners, publishers and writers can elect to have such digital rights administered in one place by a company of their choosing for a pan-European deal. You'd need to be clear on how you wanted these rights administered or if you wanted the power to remain vested in you to issue such licences as you see fit.

Consider also that if you've co-written the song your co-writers may have their own publishing structure, or be signed to a publisher. The people at Nokia will be pulling their hair out trying to get all the pieces of the jigsaw puzzle cleared. You can see why government has stepped in and asked the industry to resolve this situation.

Opportunities

- If your song doesn't succeed in one territory it has a chance in others with equally committed individuals working on it.
- The opportunity for local covers will be considered more fully in each territory as well as local synchronization use.
- There is the potential for receiving advances from each territory (how and when advances are paid are subject to in-depth discussion later), timed usually to occur on signature of contract. A number of triggers could be looked at for further advances, for example as you create success, determined by chart position, recoupment of advance or mechanical sales.

Threats

- A fragmented approach may create too much administrative chaos and extra administration costs internally. This could outweigh your income.
- Any advances received are advances against royalties yet to be earned and you need to plan for this. This is more crucial if you are representing other people's music as well as your own. You could end up in financial difficulty if success hits and you have to account to and pay the writers involved in such copyrights and haven't set money aside for this.

The societies have a huge task ahead in getting to grips with being able to act in a more proactive, entrepreneurial fashion to seize today's opportunities.

Example   **27**

Unfortunately, the collection societies are weighed down by bureaucracy and inefficient systems which waste thousands if not millions of pounds worth of writers' royalties a year. Some societies are really working hard to bring themselves into the twenty-first century, while others are terribly protectionist, especially if they are dealing with (in the main) local language and cultural music. Often, they cannot see beyond their own territorial boundaries to the bigger, worldwide picture that is at stake if the music publishing world cannot get to grips with dealing with the requests of individual cross-territory licences at society level and those controlling rights in each specific territory.

EXAMPLE

In the introduction of the book I spoke about the deals I did at my second MIDEM conference with Mike Collier. Our structure was similar to the above. The structure we chose was very deliberate and calculated to provide three things, which were at the top of our aims and objectives in building the company.

- We needed to build a strong international network and I wanted this in place with the best people in the world at that time.
- We needed some funding to help us to grow the business.
- We wanted to have potential success all over the world.

We were, in fact, probably too small for a major publisher to really consider us for a worldwide deal. If our business had been big enough I think my first deal would have still developed along a similar structure because it meant that each territory was directly responsible to us and therefore we had better communications in each territory and, as a result, developed local networks far more quickly. The structure included major publishers and major independents. Often when you do a worldwide deal, the only communication you get is in the country where you sign the deal and it becomes much harder to get a foreign office to assist. The structure that we put in place meant that every territory had to communicate with us directly. It was fantastic and opportunities flowed because of it.

Our structure was along the following lines:

- USA — Famous Music: headed by Irwin Robinson CEO in the USA and Alan Melina in the UK office. Famous was the music publishing division of Paramount films and worked to get copyrights into film and television.
- EMI — Peter Ende in the German office (fantastic support from Peter).
- The Two Peters — Peter Shoen Hoven and Peter Van Bodegarden in Benelux: terrific and supportive in all that we sought to do (later sold their company to BMG).
- EMI France also very supportive; we learnt a lot about the French market.
- Peer Australia — Matt Donlevy: we had some great hits together with Richard Branson and Virgin.

- David Gresham in South Africa: David's company is still going strong and attends the largest music conference in France (MIDEM) each year.[6]
- Fujipacific — Japan: the only subpublisher that we had a No. 1 with, via a local Japanese artist who sold well over one million singles.
- Spain and Italy were also EMI.

WORKING WITH MEDIA COMPANIES

For a long time the music industry made it as difficult as possible for media companies to gain access to content, whether it was for use in games or mobiles or other media outlets. Independent music companies with an entrepreneurial vision stepped in first, seeing immense opportunities.

Media companies need content to help fulfill customers' needs, differentiate themselves and entice consumers with simplistic solutions. The music industry wants new sources of income and routes to market to the consumer. Media companies provide these. There is significant synergy in providing this to the consumer in easy, manageable packages.

However, media companies were forced in some respects to start approaching artists, songwriters and independent record labels directly because the industry (mainly major record labels) didn't know how to deal with these companies, and what both parties wanted appeared too difficult to negotiate around. For example, I acted as business affairs/music supervisor for Platinum Sound Publishing for three years alongside A&R/music supervisor Adi Winman, and between us we managed the entire synchronization process. We were responsible for placing over 180 tracks into leading computer games and film. These leading computer games typically went to No. 1 or certainly Top 3 in the UK and European games charts. Many involved THQ and the Moto GP series. These tracks were all from new artists, new writers and mainly independent labels. We offered a 'one-stop shop' solution to the media company, providing one point of contact. We took care of the creative process of sourcing the music and all copyright clearances quickly and efficiently, securing the best fee possible. As an independent publisher we pulled the rug from underneath the feet of all UK major publishing companies, many of which then offered us their catalogues to represent! This is a good example of focusing in on one's strengths and of an entrepreneurial lead business making major waves in a short period.

There are plenty of ways for media companies to work with the music industry. Like any other business, it's about finding the right people to work with, people who are willing to be flexible and find solutions. There are many areas of commonality in business language, rights management and international development. I, and many of my colleagues, act as consultants to media companies, to assist in the development of specific projects.

---

[6]www.midem.com

Watch out for the rise of new competitors — the traditionally non-music media company. EA games has its own record label, Nokia has its own publishing division, Facebook has launched its own label, and Google is coming in fast and furious, as indicated at the BPI AGM in 2010. O2 has done deals with Live Nation and has a string of O2 academics, and of course in London the O2 Arena with AEG. The music industry has created a vast sea of competitors with routes to mass-market distribution by failing to adopt a far more proactive strategic relationship. But competition is good as it creates far more possibilities for songwriters and artists in a music industry that is changing significantly. It's a great time for music business entrepreneurs; a great time to empower business within the creative industries.

MORAL RIGHTS

'Moral rights' are an additional form of protection that protect authors and creators, rather than their economic interest right. An author or creator cannot assign or transfer his or her moral rights during their lifetime, but an author can waive these rights in a written instrument (contract). Moral rights may be waived (and used as a lever during the contract negotiation) but cannot be assigned. The three main moral rights are:

- **The right to be identified as the owner ('the paternity right')** of the right whenever a work is published, performed or communicated to the public and for this right to be upheld. The author must assert this right in writing. In layperson's language what does this mean? If you are a composer you have the right to insist that anyone using your song must credit you. The record industry doesn't like anyone to assert their rights as this provides them with a tremendous headache. So in all contracts you will be asked to waive your moral rights and then your lawyer will seek ways to bring them back in through the back door (so to speak). The record company or publisher doesn't want to give you the right to tell them that they must remove product from the shops or from digital sites because they have failed to spell your name correctly or have forgotten to credit the composers (many labels fail in this simple task), so your lawyer will add something less constraining, to say you waive your moral rights save for the right of paternity, but should Company X fail to provide the correct credit then the Company will endeavor to rectify the problem within a reasonable period or at the next print run, etc. Endeavor doesn't mean immediately and will occur only when it suits the other party.
- **The right to object to a 'derogatory' treatment of your work ('the integrity right')**. A treatment is derogatory if it distorts or mutilates a work or is otherwise prejudicial to the honor or reputation of the author. The author is the one who generally decides whether or not the use of his or her work is 'derogatory'. This can be quite subjective. Yes, a writer and artist must be protected, but at the same time a writer is engaging a publisher to find ways of placing music into synchronized use. Often, therefore, the publisher will request that the writer

stipulate any areas that they would consider unacceptable. This may perhaps be linking their music to a product they dislike or placing music into porn movies or an association with sanitary products or political ideals.

- **The right not to have a work falsely attributed to you.**

The integrity and paternity rights last as long as the copyright in the work. The right not to have a work falsely attributed to you expires 20 years after that person's death. If the author dies, the rights pass to whoever is elected in the author's will or to the person receiving the copyright. Similarly, if an error has been made, the label or publisher will be asked to change it as soon as possible, but the teeth of the right have been waived so the composer cannot insist that product must be removed from shelves and an entire release be cancelled owing to this error.

One can sympathize with labels and companies of course with genuine mistakes, but you must make sure that the teams you work with ensure that artwork is checked (build this into the contract) to minimize these errors. For most people the credit they receive is as important as the use of the music.

Moral rights exist in virtually every country, but can vary in scope from one country to another (e.g. in France, moral rights last forever), as they are not harmonized at EU level (not all member states agreeing the same terms).

US position on moral rights

The USA has never recognized a moral rights concept for music. However, there is a limited one for paintings and fine art — under American law, moral rights receive protection through judicial interpretation of several copyright, trademark, privacy and defamation statues, and through 17 USC §106A, known as the Visual Artists Rights Act of 1990 (VARA). VARA applies exclusively to visual art. Moral rights as outlined in VARA also allow an author of a visual work to avoid being associated with works that are not entirely his or her own, and to prevent the defacement of his or her works. For a historical and comparative overview of moral rights law in the USA, see Cotter (1997).[7]

If a songwriter wants protection from use then it must be specified in the writer/publisher contract so that the publisher must seek approval from the writer before use can be granted to a third party.

The Gowers Report,[8] commissioned by former UK Prime Minister Gordon Brown, did little to assist in providing support for the views of the copyright owners/creators. It provided no support to the industry's request for at least an extension in copyright to match that in the USA, where for masters it is 95 years after the first recording. This is under review and there may be a new European ruling that may see an extension to copyright, but as of December 2009 it had not been agreed. Having just received the Conservatives' round-up newsletter on the antics of the Labour government in the UK, it would appear that before the 2010 election we had two

---

[7]Thomas F. Cotter, *Pragmatism, Economics, and the Droit Moral*, 76 NCL Rev. 1 (1997).
[8]http://www.hm-treasury.gov.uk/gowers_review_index.htm

'digital ministers'. Such a shame that they didn't employ one quality person who gave a damn about the creative industries and did something to support, defend and develop the vast sums the music industry and creative industries (media) create for this country! OK, rant over. The new Conservative-led coalition may (I hope) take matters more seriously.

US COPYRIGHT ISSUES

US copyright differs from that in Europe. The Copyright Term Extension Act (CTEA) of 1998 extended copyright terms in the USA by 20 years. Since the Copyright Act of 1976, copyright would in essence last for the life of the author plus 50 years, or 75 years for a work of corporate authorship. The above Act extended these terms to the life of the author plus 70 years and for works of corporate authorship to 120 years after creation, or 95 years after publication, whichever endpoint is earlier (work for hire is common in the film industry; see Glossary). Copyright protection for works published before 1 January 1978 was increased by 20 years to a total of 95 years from their publication date. This law is also known as the Sonny Bono Copyright Term Extension Act.[9]

So now we have a 70-year term in Europe and a 95-year term on most works in the USA. UK and Europe are working hard to get an increase to a term much closer to that of the USA, but it's a slow process.

SUMMARY

Copyright exists at the point of creation, but in order to enforce your rights there must be a physical form of the copyright in existence. The copyright owner has certain moral and economic rights, but these are sometimes waived or adapted once contracts are issued and lawyers are involved. The waiving of these rights is usually requested, but a good lawyer will seek to bring these rights back into the contract (all be it in a diluted fashion) to ensure these rights are as protected as they can be and certain rights are correctly asserted. The waiving of the rights is often requested to ensure that the power within the contract is with the publisher and to ensure that the copyright owner (writer) cannot prevent any business they wish to conduct (i.e. if a record label failed to credit the writer, the writer could not come to the publisher and insist that they took action against the label and instruct all the copies to be removed from the stores).

The writer can ensure, however, that their right of paternity is respected by asserting these rights, so that if the label copy is incorrect the error is corrected on the next CD pressing, or on a digital download's meta-data, within a reasonable time-frame. The publisher, on behalf of the writer, would enforce this point. In addition, the writer can stipulate any areas of music and visual association that they

---

[9]http://www.copyright.cornell.edu/resources/publicdomain.cfm

feel are not acceptable to them or for which the publisher must seek the writer's additional approval. Such areas of concern may include an association with drugs, tobacco, animals or gratuitous violence. The writer should have the right to state whether the music can be used or not.

The thing to understand here is that the artist who has signed a contract with the record company (who owns the master) has similar rights. A film or television company must obtain clearance from both owners of the copyright (music and master). Temper this, however, with the understanding that a publisher will be trying to promote and develop income streams for the copyright work, so putting up too many obstacles may inhibit the writer's relationship with the publisher and cut back on potential income (but this is the writer's choice). If a publisher has advanced a lot of money for the copyrights then they will want to see a return on their investment. So don't be too difficult when it comes to agreeing to synchronization use.

When you start to look at and consider how you want your copyrights to be developed, ask yourself what your primary goals are. Carry out a full SWOT analysis on each opportunity and tie this in with your aims and objectives. Discuss with your lawyer, accountant or business mentor what strategy best fits with your stated position.

DIANE WARREN – WRITER

At Musexpo 2008, Diane Warren, who has written over 2500 songs and has her own publishing company, Real Songs,[10] employing her own staff, obtaining her own covers and synchronizations, spoke about the most fortuitous turn of events in her life. Her first publishing deal some 25 years ago ended in court, and as a result her new songs (copyrights) had to be administered. Her lawyer at this time advised her to administer herself until all legal actions were resolved. This she did and in the intervening months had a huge hit with Laura Brannigan; Real Songs was born. This time she sought to self-publish, and now only publishes her own works; she does not represent other writers. Her hits include:

- Celine Dion – 'Because You Loved Me'
- Toni Braxton – 'Unbreak My Heart'
- LeAnn Rimes – 'How Do I Live'
- Aerosmith – 'I Don't Want To Miss A Thing'
- Cher – 'If I Could Turn Back Time'
- Faith Hill – 'There You'll Be'.

The wonderful thing about the industry as it stands today is that it is possible to do so much more yourself or as much or as little of that as you want or need. It is the time of the independent and hard working creatives.

Diane Warren

[10]www.realsongs.com

Managing rights — international framework and issues

3

I have to share a short story that deeply touched me. I had the pleasure of visiting New Orleans for Jazz Fest 2008 and was really taken aback by the people I met and the incredible musicianship I encountered. I cried, sang and danced with the folks of this wonderful town in the aftermath of hurricane Katrina.

You will recall the thousands of people displaced after the hurricane. New Orleans is regarded as being the home of jazz music in America.[1] Some people lost everything, literally everything — some people lost their lives. My sister was caught up in this; the school she worked in was all but swept away. The US government largely forgot these unfortunate people and many people in the rest of America failed to speak up for them. Slowly (very slowly), amenities were restored and life in some way got back to normal, but this took months and years, not weeks. In addition to the personal tragedy that the hurricane brought it tore apart the home of this great music city. For those who have never been to New Orleans, you must go! The 2008 Jazz Fest saw the return to New Orleans of the Neville family, the first time they had returned to New Orleans since Katrina, having suffered great tragedy.

So on this very day the Neville brothers and their family returned to New Orleans and people came out to welcome them back. The local people stood shoulder to shoulder with this great family to share their sorrow and pain. Charmaine Neville (daughter and niece to the Neville brothers) kicked off the day and blew the storm clouds away. What a lady; newly composed songs struck a real chord with those listening. Several other fantastic local artists followed in the gospel tent. Just when you thought things couldn't get any better along came Aaron Neville. Aaron had buried his wife just weeks before, and was clearly tormented with pain. In addition, this gig was his first back among his people in his hometown. Every note he sung was racked with emotion. Every lyric appeared to pierce his heart. We cried with him, we applauded his courage, and his wonderful voice and band. He was given a standing ovation, which brought the house down, forcing him to come and do two more encores. To Aaron Neville, thank you for this moment. Your return to New Orleans will never be forgotten and I felt so privileged to be a tiny part of this historic day. The people loved you for your courage and your wonderful music.

Thinking the day just couldn't get any better; I strolled over to the jazz tent and saw Chick Corea (wow), and then to the surprise of the day: Irvin Mayfield and the

---

[1]http://www.neworleansonline.com/neworleans/music/

The Art of Music Publishing. DOI: 10.1016/B978-0-240-52235-7.10003-X

New Orleans Jazz Orchestra. Stop right now and go and Google them or go to their website.[2] Oh, my word ... just incredible, the musicianship was stellar, beyond brilliant. I will return and be happy just to see them. They were just awesome. I wasn't the only one who thought that. Rubbing shoulders with Jude Law (swoon ... sorry, chaps), it was clear he too felt the same. NOJO started their world tour 2009/2010, and have been privately funded by a fellow American, and embraced by the good people of New Orleans. When the government let the people of New Orleans down, individuals stepped up and did what they could. This is the result of one such intervention. I make no apologies for veering off subject, this is a music book after all, and I intend to share as much with you as I can. It starts and finishes with music, this is why I'm writing this book, and this is why you're reading it. Before leaving New Orleans, I stopped by 'Preservation Hall',[3] a room as tiny as can be. It only seats about 50 people at one time and is dedicated to the love of music and respect of tradition; you'll find all the greats have played here, including Louis Armstrong. Preservation Hall now happens to be run by a friend of the family (I didn't know this until then).

OK, so now we'll go back to the chapter and I'll relate the relevance of this content to all the music I heard in New Orleans that year.

You'll begin to appreciate that no matter where you are in the world, songwriters have to be able to collect their income because this is how they live. We're going to take a look at how publishers achieve this for them. How is money moved around the world to get back to you in your home territory, wherever it is?

COLLECTION SOCIETIES

The world of music publishing is structured with an interwoven network of collection societies. They break down into societies that collect different parts of the income stream to composers and publishers, namely income from performing rights (radio, television, film, touring) and mechanical income payable by record labels from the sale of recordings (digital, physical, streaming) containing the copyrighted works of songwriters and publishers. Their job is not just to represent local rights but to provide reciprocal protection and collection of everyone's rights when they are activated in their territory.

In other words, there is an integrated network where being a member of one society provides a common set of principles around the world for the protection of and the collection of royalties.

Collection societies have existed since 1852 with the birth of SACEM, the French Performing Right Society. MCPS and PRS (UK) were established in 1911 and 1914, respectively. The two UK societies have merged in recent years to reduce costs and find efficiencies for the benefit of its members. The UK society is now called PRS for Music.[4]

---

[2]www.thenojo.com
[3]www.preservationhall.com
[4]http://www.prsformusic.com

In the USA, three performing rights societies exist, which essentially do the same job, but have different characteristics. ASCAP is run and controlled by writer members and BMI is run and controlled by publisher members. SESAC grew primarily from looking after central European clients. All are proactive in supporting new composers and can assist worldwide in this regard. All have a proactive stance on education and sharing information on performing rights. Here's some information about each of them. Visit their websites and go and see their representatives in whatever country you are based — be proactive and find out more for yourself.

BMI

BMI does not charge writer members a fee to join, but there is currently (April 2010) a fee of US $150 if you wish to start an individually owned music publishing company through them, which rises to $250 for a corporation.

Formed in 1939 as a non-profit-making performing right organization, BMI was the first to offer representation to songwriters of blues, country, jazz, R&B, gospel, folk, Latin and, ultimately, rock 'n' roll.

BMI was founded by broadcasters to provide competition in the field of performing rights, specifically the songwriters and composers who were disenfranchised by the existing system. BMI was created to assure royalty payments to writers and publishers of music who were not represented by the existing performing rights organization, and to provide an alternative source of licensing for all music users.

BMI's history coincides with one of the most vibrant, evolving and challenging periods in music history. As popular music has moved from big band to rock 'n' roll and hip hop, and formats have evolved from 78 and $33\frac{1}{3}$ rpm vinyl records to compact discs, MP3s and beyond, BMI has worked on behalf of its members to maintain a leadership position not only in the USA, but worldwide.

"Underlying everything BMI does is its philosophy; an open-door policy that welcomes songwriters, composers and music publishers of all disciplines, and helps them develop both the creative and business skills crucial to a career in music" (BMI, 2009).[5]

PHIL GRAHAM – SENIOR VICE-PRESIDENT, WRITER/PUBLISHER RELATIONS AT BMI

At an international music convention (January 2010) I met up with Phil Graham, Senior Vice-President of Writer/Publisher Relations at BMI, New York. I asked Phil for insight into various points of interest. Here are some of the points we discussed, along with his answers.

Q: What do you consider the biggest challenge at the moment for the industry?
"That would be the current economic challenge we all face. It impacts everything; even before we look at the effects of rights issues for the music industry in the digital arena.

[5]http://www.bmi.com (members include Eric Clapton, Elton John, Fall Out Boy, Daniel Powter, Colbie Caillat, Christina Aguilera, Josh Groban, Jennifer Lopez and many more).

Mechanicals have flattened out, but at the moment flat is the new up. It's not just down due to the economy, but because of the challenge of licensing digital media in all formats – YouTube, Spotify, MySpace, Second Life. We're actively trying to licence the use of music in all new digital areas that develop".

Q: Do you feel the blanket ISP licence may be the way to go?

"I think it could be an efficient way to do it. It would be easy to administer and it could remove the bad press the industry has received when we have to take action to protect our songwriters' interests. It would allow us all to deal with the bigger issue; focus on the massive task of documentation of the works used and the distribution of licence fees that result from the deals that are struck on behalf of the songwriter and music publisher. There are challenges with all digital media, it's very complex, and at present there needs to be a better fit for consumers and businesses …. I believe the all-you-can-eat model of streaming service may be a key model moving forward. Personally it would suit how I would consume music as an individual and may be a very effective route for the industry".

Q: Would the US industry benefit from a strategic lobbying alliance on fundamental issues? In the UK we now have a collective voice under 'UK music'.

"It's interesting to see that the UK is doing this. In America we work on the issues directly and we get the best available agreements we can and deal with hundreds of new licences a year …. In the USA Fred Cannon deals with legislative lobbying, at the state and government level, for BMI".

Q: What initiatives have BMI got running for new and developing writers in the USA and UK?

"BMI has always been very active in assisting writers. We put on or are involved with hundreds of showcases featuring new talent every year. We present musical theater workshops, TV and film scoring workshops, even US music scholarships, and UK scholarships with the commercial music department at the Royal Academy of Music. We constantly interact with and for songwriters and composers when their careers and talents are ready.

"We also work with American writers going to the UK, and we help PRS writers whom BMI collect for in the USA when they organize trips to Nashville, Los Angeles and the like. On both sides of the ocean we provide guest speakers to educate new writers on the importance of performing rights and what BMI does".

Q: How does BMI help in new countries developing inroads into copyright structure?

"We play a role in helping develop new territories to become more receptive to copyright. We provide technical visits and information exchanges to help educate where possible. We have a dedicated team for this".

Background

As Senior Vice-President, Writer/Publisher Relations at BMI, Phillip R. Graham directs and oversees the writer/publisher activities in BMI's seven offices: New York, Los Angeles, Nashville, Atlanta, Miami, London and San Juan. Graham's staff work to support and advance the careers of songwriters, composers and publishers who licence their works through BMI.

Before working with BMI, Graham worked for CBS Records in Nashville, as well as with retail and record manufacturing companies. After joining BMI in 1979, he worked for seven years in BMI's Nashville Writer/Publisher Relations department. He then relocated to London in 1987 as Director of European Writer/Publisher Relations. He was named Vice-President in 1991 and Senior Vice-President in 2003. Later in the year, he moved to BMI's headquarters in New York City, where he currently is based.

A native of Evansville, IN, he received a Bachelor's degree in business administration from Vanderbilt University in Nashville, TN.

ASCAP

There is a writer or publisher joining fee of US $25 as at 2010.

"ASCAP is a membership association of more than 350,000 US composers, songwriters, lyricists, and music publishers of every kind of music. Through agreements with affiliated international societies, ASCAP also represents hundreds of thousands of music creators worldwide. ASCAP is the only US performing rights organization created and controlled by composers, songwriters and music publishers, with a Board of Directors elected by and from the membership" (ASCAP, 2009).[6]

SESAC

"SESAC was established in 1930 and built on service, tradition and innovation. SESAC was founded in New York in 1930 by German immigrant Paul Heinecke, who, in an effort to help European publishers with their American performance royalties, established SESAC as the Society of European Stage Authors and Composers. Throughout the decades, until his death in 1972, Paul Heinecke guided SESAC with his own unique mix of old-world charm and 20th century savvy. With an established cornerstone repertory of the finest European Classical Music, SESAC began to turn its attention to American music in the 1930s. Today, however, the company is known simply as SESAC. With an international reach and a vast repertory that spans virtually every genre of music, SESAC is the fastest growing and most technologically adept of the nation's performing rights companies".[7]

Harry Fox

Harry Fox is the mechanical rights agency (collecting monies due to writers and publishers from the sale of each record, physical and digital) in the USA. Here's a little bit about them.

"The Harry Fox Agency represents music publishers for their mechanical and digital licensing needs. We issue licences and collect and distribute royalties on our affiliated publisher's behalf. This includes licensing for the recording and repro-duction of CDs, ringtones, and internet downloads. HFA no longer issues synchronization licences for the use of music in advertising, movies, music videos, and television programs, but we do collect and distribute on synchronization licences that were granted prior to our discontinuation of synchronization service in 2002. HFA also conducts royalty examinations, investigates and negotiates new business opportunities, and pursues piracy claims".[8]

---

[6]http://www.ascap.com (Duke Ellington, Dave Matthews, George Gershwin, Stevie Wonder, Leonard Bernstein, Beyoncé, Marc Anthony, Alan Jackson, Henry Mancini, Howard Shore and many more).
[7]http://www.sesac.com/About/About.aspx
[8]http://www.harryfox.com

So these are the organizations to be aware of in the USA. Now to the picture of what happens in the UK and Europe.

UK PERFORMING AND MECHANICAL RIGHTS SOCIETIES
Writer and publisher membership details of PRS for Music[9]

Up until recently and in all books to date you will have read that there are two UK societies, the PRS (Performing Rights Society) and MCPS (Mechanical Copyright Protection Society). In recent times they have come together under one umbrella to share their database and common infrastructure. They have amalgamated under a common name PRS for Music, but within this structure still carry out specific functions. PRS collects performance income (the US equivalent being BMI, AAS-CAP and SESAC) and the MCPS (the US equivalent being Harry Fox) collects mechanical income. It is likely that this structure may continue to develop for reasons of ergonomics. I will discuss a common umbrella under PRS for Music with two distinct roles as they are at present.

A writer or a publisher must meet certain criteria to join the MCPS or PRS (or both).

PRS for Music (administering performing and mechanical rights) — requirements 2010

In order to become a PRS writer member there is a small fee of £10 payable (2010).
You must have had one piece of music that has been:

- broadcast on radio/television
- used online
- performed live in concert
- otherwise played in public.

Only one of these criteria needs to be met, but evidence must be provided to support your application.

In order to join as a PRS publisher member you will need to represent at least 15 pieces of active music. Once again, evidence of this will need to be provided upon registration. There is a fee of £400 (2010) for joining. Consult the relevant websites for changes in membership criteria.

MCPS — requirements 2010

In order to join the MCPS as either a writer or publisher you must have written one piece of music, or your catalogue must contain one piece of music that has been:

- commercially released by a record company (other than a record company you own)
- recorded into a radio or television program

---

[9]http://www.prsformusic.com

- recorded in an audiovisual or multimedia production
- used online.

For a writer to meet the above criteria it is also important that the piece of music is not contracted to a music publisher. Evidence of the above uses must be provided upon registration.

Benefits of joining
Joining the PRS for Music carries with it many benefits, the majority of which are listed below:

- fast and efficient collection and payment of your UK and international royalties
- quarterly royalty payments on *all* revenue streams
- accurate radio and television analysis, where royalties are attributed to actual performances
- regular members' workshops and networking sessions
- an online services package
- direct access to broadcaster logs, ensuring accurate payments for radio and television usage, using the latest tracking technologies to identify the rest
- access to policies and procedures, career support, administration rates, payment information, etc.

PRS for Music:

- works with UK and European Union (EU) governments to improve the rights of creators' endeavors to be at the forefront of licensing new platforms in Europe including online, podcasts, ringtones, mobile TV, streaming and all other new uses as they develop
- can represent rights holders worldwide, as they have reciprocal arrangements with similar societies in each foreign territory.

Go over to the 'PRS for Music' website and read some of the fantastic research work and papers put together by Will Page, chief economist at PRS for Music in the UK. You can subscribe to the news feeds and download insightful papers on leading topics in the press. As the rate of change is substantial I'm not going to pick on any particular topic to highlight but I will say that one of Will's reports in 2009 highlighted the fact that expectations moving forward to 2011 are that lost revenue (based on reducing mechanical income) is likely to be bridged with alternative income revenues, and for music publishers that's got to be great news! Not so good for major record labels when the music industry's global sales have fallen by 30% over the last five years, even though digital sales grew by 940% in that time, according to the IFPI. It estimates that, overall, music sales fell by 10% in 2009 to $15.8 billion.

To give you a small snapshot of the level of income collected in the UK:

- broadcast and online: 180.2 million
- international: 139.8 million

- public performance: 146.6 million
- recorded media: 141.6 million

Broadcast and online can be broken down further to:

- television: 105.1 million
- radio: 51.8 million
- online: 17.6 million
- ringtones: 5.7 million.

A list of societies around the world

- USA: BMI — performance income (www.bmi.com)
- USA: ASCAP — performance income (www.ascap.com)
- USA: Harry Fox — mechanical income (www.harryfox.com)
- USA: SESAC — performance income (http://www.sesac.com)
- Australia: APRA — performance income (www.apra.com.au)
- Australia: AMCOS — mechanical income (www.amcos.com.au)
- Belgium: SABEM — performance and mechanical income (www.sabam.be)
- Canada: SOCAN — performance income (www.socan.ca)
- Canada: CMRRA — mechanical income (www.cmrra.ca)
- Canada: SODRAC — mechanical income (www.sodrac.com)
- Denmark: KODA — performance income (www.koda.dk)
- Denmark: NCB — mechanical income (www.ncb.dk)
- Finland: TEOSTO — performance income (www.teosto.fi)
- Finland: NCB — mechanical income (www.ncb.dk)
- France: SACEM — performance and mechanical income (www.sacem.fr)
- Germany: GEMA — performance and mechanical income (www.gema.de)
- Iceland: STEF — performance income (www.stef.is)
- Iceland: NCB — mechanical income (www.ncb.dk)
- Ireland: IMRO — performance income (www.imro.ie)
- Ireland: PPI — mechanical income (www.ppiltd.com)
- Italy: SIAE — performance income (www.siae.it)
- Italy: SCF — mechanical income (www.scfitalia.it)
- Japan: JASRAC — performance and mechanical income (www.jasrac.or.jp)
- Netherlands: BUMA — performance income (www.bumastemra.nl)
- Netherlands: CEDAR — performance income (www.cedar.nl)
- Netherlands: STEMRA — mechanical income (www.bumastemra.nl)
- Norway: TONO — performance income (www.tono.no)
- Norway: NCB — mechanical income (www.ncb.dk)
- Portugal: SPA — performance income (www.spautores.pt)
- South Africa: SAMRO — performance income (www.samro.org.za)
- South Africa: SARRAL — mechanical income (www.sarral.org.za)
- Spain: SGAE — performance income (www.sgae.es)
- Sweden: STIM — performance income (www.stim.se)

- Sweden: NCB — mechanical income (www.ncb.dk)
- Switzerland: SUISA — performance and mechanical income (www.suisa.ch)

The main advantage of this network is the bargaining power of such societies when they work together, lobbying and assisting in the development of legal support structures for the ever-changing permissions and royalty payments for new income streams.

To clarify: the network of societies and the reciprocal agreements between member societies allow copyrights to be administered all over the world.

THE BERNE CONVENTION

So now you have a basic understanding of how a UK songwriter receives royalties from Spain, or an American writer receives royalties from Australia. When you join PRS for Music in the UK, your fellow Frenchman has a similar society in France (SACEM) and the societies have an international agreement for the protection of your royalties and collection of income. However, the harmonization of these copyright issues has been achieved via the international convention of copyright, which came about at a meeting in Berne in 1886 (now referred to as the Berne Convention). There were 14 original member states, which as of 2010 had grown to 184. Since almost all nations are members of the World Trade Organization, the Agreement on Trade-Related Aspects of Intellectual Property Rights requires non-members to accept almost all of the conditions of the Berne Convention. The core of the Convention provides for each contracting country to provide automatic protection for works in other countries of the union and for unpublished works whose authors are citizens of or residents in such other countries.

Appreciating that the common laws in each country are different from one country to another, the various conventions on copyright were set up to provide common goals and understanding to which each country agreed, thus creating a formidable international network of common structure and rules.

The UK is a member of several international conventions, notably:

- **the Berne Convention** — for the protection of literary and artist work
- **the Rome Convention** — for the protection of performers, producers of phonograms and broadcast organizations.

Both conventions (and others outside our industry) are administered by the World Intellectual Property Organization (WIPO). UK nationals or residents falling within the scope of one of these conventions are automatically protected in each member country of the convention by the national law of that country.

So now we have a road map of copyright treaties around the world to which members will sign up, to help unify areas of copyright law and understanding and to allow the flow of business.

Note: Most treaties will require that copyright owners identify their work by using an international copyright symbol: ©. Berne does not insist on this.

COPYRIGHT ISSUES LEADING EUROPEAN DISCUSSION

There are several hot topics currently affecting the ability of various structures in society to operate, and the development of competition between societies, and which the EU is determined must be tackled in some way. In simple language, I will describe the issues and refer to various well-written articles on the topic. The website will assist with ongoing content.

Technological developments and rights need to be cleared for use across territorial boundaries quickly and efficiently. Imagine you are a Tel Co and you want a track for a European television commercial and then to offer this as a free or reduced download via your mobile network. That Tel Co would potentially have to seek agreement in every country it wished to use the copyright for, with a differing cost in each country and a contract with each one. Then it would need a licence from the owners of the master recording (the label); in addition, the sale of the download or free download must include a mechanical fee payable to the publisher in each territory collected by the relevant society.

The EU Commission has insisted that this 'closed shop' is something approaching a cartel rather than a tried and tested collection system of complicated copyright issues. They argue that such a system is preventing business from flourishing and that the industry must do more to assist in such areas.

The door has been kicked open by Brussels to bring competition into the market. Easier said than done! So now a major label can shop around to obtain the best mechanical fee on offer for a European-wide licence from any society it wishes (in theory), pitching society against society.

In addition, a publisher can elect to give any society it wishes the right to collect its digital rights. So instead of the Tel Co having to pay each society in Europe it can now (in theory) obtain one licence, at a negotiated rate for a European deal.

However, it's not as simple as this. No single society wants to lose business, and various court actions are being tried (society fighting society). The societies representing more popular works of international standing are likely to do better as their catalogues are going to be in greater demand and therefore their influence is more significant.

Where could this lead?

- It's early days, but moves are afoot already with the rise of 'super societies'. (Watch this space.)
- Central licensing — its importance is increasing.
- Pan-European licensing — its importance is increasing.
- Internet service providers (ISPs) — differing territorial perspectives.

At the BPI AGM in 2010, Google announced the launch of their label and made it very clear to the industry it will not accept that 23-plus contracts are needed to do a worldwide deal for the licensing of a single track. They want one deal for the master, one deal for the music. This has got to be the industry's biggest challenge. If we cannot appease such a customer and serious competitor (they are massing), then we can expect the industry to be further damaged. These 'super competitors' are a growing and very serious threat.

IMPLEMENTATION OF EU DIRECTIVE — UK COPYRIGHT ACT

Copyrights owners will now be able to obtain injunctions against ISPs, who are often in the middle between rights owners and infringers, providing the ISP actually knows "that the service is being used to infringe copyright". The copyright owner may have to serve notice on the ISP. This should start to provide an effective means of stopping copyright infringers.

A new broad 'right of communication' to the public for copyright owners came into force in March 2001, covering communication to the public by all methods of electronic transmission.

All works, literary, music, photographs that are considered to be 'freely available' on the internet are still covered by copyright law. Just because the medium is different and allows for a mass circulation of information, it does not mean that they are 'free'. All owners of copyright works are also entitled to have their moral rights upheld, i.e. paternity, integrity and false attribution.

A new 'technology-neutral' right of communication to the public has been introduced, which supersedes the previous rights to broadcast and transmit to subscribers via a diffusion service. This 'right of communication' for copyright owners covers sending material electronically, for instance by e-mail, and making material available online.

WIPO

The World Intellectual Property Organization (WIPO) is a "specialized agency of the United Nations"[10] set up to "develop a balanced and accessible international intellectual property system".[11]

WIPO was established in 1970 and currently has 184 member states. It provides its members with a forum to "negotiate intellectual property treaties and standards"[12] and assists "governments in using intellectual property" and developing "the policies, structures and skills needed to harness the potential of IP for economic development".[13] Its headquarters are in Geneva, Switzerland.[14]

---

[10]http://www.wipo.int/about-wipo/en/what_is_wipo.html

[11]http://www.wipo.int/about-wipo/en/what_is_wipo.html

[12]http://www.wipo.int/freepublications/en/general/1007/wipo_pub_1007.pdf

[13]http://www.wipo.int/freepublications/en/general/1007/wipo_pub_1007.pdf

[14]http://www.researchtheworld.com

ISP NEWS
UK

In 2009 there was an important change in the attitude of governments towards accepting "that the carriers of digital content must play a responsible role in curbing"[15] music piracy. European governments have been the most active in pursuing this role. This led in 2010 to the new Digital Economy Bill. In short, this is legislation to tackle internet piracy, including bans for illegal file-sharers. The Digital Economy Bill and how it is implemented have yet to be measured as a success or failure. Peers had earlier rejected a bid by ministers to include wide-ranging powers over future online piracy law.

The UK government has also made its position clear: "Internet Service Providers (ISPs) will be legally required to take action against any of their clients who access pirated material".[16]

This has got to be one of the hottest topics to monitor and we will have to wait and see what develops from here, and how it will be applied.

Europe

In France we saw the publication of the Sarkozy Agreement, requiring ISPs to disconnect serial copyright infringers. This report has also been backed up by a court in Belgium, where the judgment in the Sabem–Tiscali case confirmed, "ISPs must take responsibility for curbing infringements on [their] networks".[17] Sweden has also made its stance clear through the Renfers Report, which suggested that a change in the law should take place which obliges ISPs to take action in terminating the contract of repeat offenders.

Japan

Although Japan has seen a great deal of growth in legal downloads, it also has to confront an ever-growing problem with music piracy. Consequently, the Recording Industry Association of Japan is launching the largest ever education campaign with regard to mobile piracy, and has secured support from the government, mobile operators and music rights holder organizations. Personally I think education may be a stronger tool than we realize, especially if it starts young, while at school.

USA

The USA has not seen the same progress as in Europe or Japan; however, AT&T has acknowledged the role of ISPs and their responsibility with regard to dealing with large-scale copyright infringement.

---

[15]http://www.ifpi.org/content/library/DMR2008.pdf
[16]http://www.musicweek.com/comments.asp?storycode=1033165
[17]http://www.ifpi.org/content/library/DMR2008.pdf

In a 2010 article, the US government stated that iPhone owners are officially allowed to 'jailbreak' their devices for 'educational purposes'. This rule was one of a number of exemptions to 1998's Digital Millennium Copyright Act (DMCA) anti-circumvention protections. These exemptions are reviewed and authorized every three years to ensure that work protected by copyright can be used in non-infringing ways. While the 'jailbreaking' exemption is surprising, there is a list of others to make note of:

- to bypass video game protections in order to investigate or correct security flaws
- for college professors, film students and documentary filmmakers to break the copy-protect measures on DVDs for embedding clips for educational purposes, criticism or in non-commercial videos
- to bypass the use of external dongles if the dongle no longer works
- for used cellphone owners to unlock phones so that they can be switched to another wireless carrier (this was a renewal of a 2006 exemption).

The Electronic Frontier Foundation (EFF), which filed for three of these exemptions, is understandably ecstatic.[18]

Sam Gustin for Wired (http://www.wired.com/epicenter/2010/11/coica-web-censorship-bill/) is reported as saying:

> *On Thursday, the Senate Judiciary Committee unanimously approved a bill that would give the Attorney General the right to shut down websites with a court order if copyright infringement is deemed "central to the activity" of the site — regardless if the website has actually committed a crime. The Combating Online Infringement and Counterfeits Act (COICA) is among the most draconian laws ever considered to combat digital piracy, and contains what some have called the "nuclear option," which would essentially allow the Attorney General to turn suspected websites "off."*
>
> *COICA is the latest effort by Hollywood, the recording industry and the big media companies to stem the tidal wave of internet file sharing that has upended those industries and, they claim, cost them tens of billions of dollars over the last decade.*
>
> *The content companies have tried suing college students, They've tried suing internet startups. Now they want the federal government to act as their private security agents, policing the internet for suspected pirates before making them walk the digital plank.*
>
> *Many people opposed to the bill agree in principle with its aims: Illegal music piracy is, well, illegal, and should be stopped. Musicians, artists and content creators should be compensated for their work. But the law's critics do not believe that giving the federal government the right to shut down websites at will based upon a vague and arbitrary standard of evidence, even if no law-breaking has been proved, is a particularly good idea. COICA must still be*

---

[18]http://mashable.com/2010/07/26/dmca-jailbreak-exemption/

approved by the full House and Senate before becoming law. A vote is unlikely before the New Year.

This is a significant action, one that will reverberate and shake companies to the core. It's a bit like using a sledge hammer to crack a walnut however, But where reasoned argument is replaced with provocation, and the right to earn a living is being taken away from composers and musicians I wonder if such people or companies would feel the same if anyone could just take money from them 'just because the could'. Just because technology 'allows' theft to take place, isn't it still wrong? Therefore what is wrong in protecting such people? When exactly did technologists get a free passport to crime? Young misguided people I speak to still say well, if I can get it for nothing, why should I pay for it? Isn't that the difference between knowing right from wrong? Theft is never right, no matter what the circumstance. Because you've left your window open does that given permission to a burglar to rob you? So why is it different with copyrights?

CENTRAL LICENSING

Moving on now, I want to discuss the issue of central licensing. I came across the following article, which I have referenced accordingly, but which provides good detail on what is a complicated and developing issue.

COMPETITION: COMMISSION RENDERS COMMITMENTS BY MUSIC PUBLISHERS AND COLLECTING SOCIETIES LEGALLY BINDING (EUROPA, 2006)

The European Commission has made legally binding under EC Treaty competition rules the commitments given by the five major music publishers (BMG, EMI, Sony, Universal and Warner) and 13 European collecting societies (AEPI, AustroMechana, GEMA, MCPS, MCPSI, NCB, SABAM, SDRM, SGAE, SIAE, SPA, STEMRA, SUISA), the signatories of the Cannes Extension Agreement, regarding Central Licensing Agreements. The commitments ensure that record producers can continue to receive rebates from collecting societies on royalties paid in the context of Central Licensing Agreements. These rebates are currently the only form of price competition among collecting societies. The commitments also ensure that potential entry by collecting societies in the music publishing or record production markets is not impeded. The Commission had been concerned that two clauses of the Cannes Extension Agreement may have violated the EC Treaty's ban on cartels and restrictive business practices (Article 81) but has now closed the case in the light of the commitments.

"Under a Central Licensing Agreement, a record company can obtain a copyright licence for the combined repertoires of all the collecting societies and covering the whole of the EEA or part thereof, from any collecting society within the EEA. Central Licensing Agreements are an example of how competition among collecting societies for the granting of pan-European licences can function, to the benefit of all involved.

"The Cannes Extension Agreement is an agreement between 13 European collecting societies managing mechanical copyright (the right involved in the production of physical carriers of sound recordings, such as CDs) and the five major music publishers, which are members of these societies. The Agreement settles a number of issues regarding the relations between the two groups of companies.

"The commitments offered by the parties to the Agreement concern two clauses of the Agreement on which the Commission had expressed its concerns. The first commitment ensures that collecting societies may continue, in the context of Central Licensing Agreements, to give rebates to record companies, paid out of the administration fees that they retain from the royalties which they collect on behalf of their members. Rebates are currently the only element of price competition in this market. The second commitment consists in the removal of a no-competition clause, which would have prevented collecting societies from ever entering either the music publishing or the record production market. The Commission decision, based on Article 9 of the procedural Regulation 1/2003 on the implementation of the EC Treaty's competition rules, takes into account the outcome of consultations on the commitments offered by the parties to the Agreement. This decision ends the proceedings concerning the Cannes Extension Agreement. However, if the parties to the Agreement were to break their commitments, the Commission could impose a fine of up to 10% of their total turnover without having to prove any violation of the EC Treaty's competition rules".[19]

[19]http://europa.eu/rapid/pressReleasesAction.do?reference=IP/06/1311&guiLanguage=en

INTERNET LICENSING — GROWTH OF LEGITIMATE ONLINE LICENSING

A recent study by the EU Commission concluded that the main obstacle to the growth of legitimate online content services in the EU is the difficulty in securing attractive content for online exploitation. In particular, the present structures for cross-border collective management of music copyright — which were developed for the analogue environment — prevent music from fulfilling its unique potential as a driver for online content services. The Commission proposes options to remedy this situation as only music has the real potential to kickstart online content services in Europe in line with the Lisbon agenda.

Internal Market and Services Commissioner Charlie McCreevy said: "We have to improve the licensing of music copyright on the internet. The absence of pan-European copyright licences makes it difficult for new European-based online services to take off. This is why we are proposing the creation of Europe-wide copyrights clearance. Central clearance is not about making content available on the cheap. It offers a model whereby Europe's creative community will get the lion's share in revenues achieved online".

The study examined the present structures for cross-border collective management of copyright for the provision of online music services. It concluded that the absence of EU-wide copyright licences for online content services makes it difficult for these music services to really develop and grow. Online music services targeted by the analysis include services provided on the internet — such as simulcasting, webcasting, streaming, downloading or an online 'on-demand' service — and also music services provided to mobile telephones. The study focuses on these services because all of them can be enjoyed across Europe and, as a consequence, their copyright needs to be cleared throughout Europe.

It concludes that entirely new structures for cross-border collective management of copyright are required, and that the most effective model for achieving this is to enable right-holders to authorize a collecting society of their choice to manage their works across the entire EU. This would create a competitive environment for cross-border management of copyright and considerably enhance right-holders' earning potential.

In addition, the right-holder's freedom to choose any collecting society in the EU would create a powerful incentive for these societies to provide optimal services to all their right-holders, irrespective of their location, thereby enhancing cross-border royalty payments.

Possible outcomes of these unresolved issues

- Media customers want a fast, effective, single rate to licence works for use in Europe.
- The industry has no effective charter to do this and should one be made it could fall on its own sword and be accused of acting as a cartel.
- Societies will be forced to compete with each other, causing in-fighting, frustration and a negative outcome to the customer, who wants a fast and effective solution to licensing music.
- This new area of concern should not be underestimated in its importance or impact. It will be seen as yet another reason for the media industries to look for and develop their own solutions and routes to market. What do I mean by this? EA Games is setting up its own record label and signing artists or projects. The Tel Cos are all following suit in setting up publishing companies and enticing artists and projects out of contract to sign or develop music for them. As these companies develop greater distribution channels and direct customer delivery it is inevitable that growth in this highly contentious artist product development will grow. For this reason alone the music industry cannot sit and fight it out as opportunity after opportunity slips through its hands.

I would like to remind you that a business will only flourish and grow if it is fulfilling the needs of its customers. The music industry single-handedly sued its customers when it failed to act on internet piracy. It put up obstacle after obstacle when media companies needed content for licensing. It also missed the opportunity to deliver legitimate download sites and allowed Apple (a non-music company) to become the largest digital provider in the world. Apple now has direct marketing capabilities greater than all the major labels.

The issue of central licensing is in itself fairly insignificant but could just be the straw that breaks the camel's back. If left unresolved it could be a significant catalyst in driving media industries to find their own solutions or swallowing up the industry as it stands, or more likely picking off the artists (brands) they can commission and own. Those that have effective distribution channels to consumers will be dominant (who will be the slave, who will be the master?). As physical product is replaced by digital, as high-street stores are replaced by digital stores and lockers (in the Cloud), the music industry will go head to head with companies who have only ever known the digital space!

LEGAL COMMENT FROM ANTHONY HALL – A DUAL-QUALIFIED LAWYER

(Eleven years PQE English solicitor and admitted to the Bar as a New York Attorney in 2002)

Should 'big media' keep suing their customers and now their partners too?

Copyright is a fairly recent addition to modern legal systems. Whereas basic 'human rights' began to be enshrined in the development of society fairly early on [e.g. certain ancient Sumerian laws, even the Ten Commandments (in places), Magna Carta], copyright, as a codified part of the law, has evolved essentially (and almost always) in response to the development of technology. Widespread adoption of Caxton's printing press led Henry VIII to set up the 'Stationers' Company' which in due course led to the Statute of Anne, the world's first dedicated copyright legislation in 1709 (a mere 300 years ago!).

The twenty-first century – and the explosion of incredible and cheap technology into our homes which facilitates the easy reproduction of words, music, lyrics, photos and many other forms of IP – brings a whole new urgency to the debate surrounding copyright and whether our legislation, to coin an overused phrase, is currently 'fit for purpose'.

Copyright, in certain areas, e.g. terms and types of protection, has evolved considerably; for example, from the 14 year's protection in the 'printing right' set out in the Statute of Anne, literary copyrights are now afforded protection for 70 years after the author's death. Copyright now additionally protects a wide variety of intellectual property (IP); including sound recordings, choreography and even computer programs. Over the past two years, the recording industry in Europe has been lobbying in the EU to increase the term of copyright for sound recordings from 50 years to perhaps 70 or 95 years, and thus bring it more into line with the term of protection in the USA. Other lobbying groups, such as EFF (the Electronic Frontier Foundation) and the Pirate Party (an offshoot of Pirate Bay), have gone on record to seek and petition for a reduction in copyright term, and even for the abolition of copyright altogether in certain circumstances. In recent years, a separate form of copyright (and related rights) called Creative Commons has also come into existence, although it is not widely used outside academic circles. As the Chinese curse so succinctly states in the world of copyright … "We live in interesting times".

The biggest challenge to copyright over the past 10 years or so has arisen from our use of the internet or worldwide web. As bandwidth has increased, from the 'slow as a snail's pace' dial-up services in the 1990s to the superfast optical broadband many of us enjoy using today, our ability 'as consumers' to copy, use, enjoy and exploit others' copyright material (and IP generally), often without payment, has exponentially increased. The major record companies (and Metallica) put Napster out of business in 2000/2001, but the so-called 'pirates' simply moved their services elsewhere; to Limewire, to P2P generally, to Pirate Bay. With every RIAA writ against file-sharers ('a finger in the dike?'), another two or three (million) appear, threatening to drown out all copyright owners in the deluge.

So what are the solutions? Currently a number of similar propositions are being tested in different territories.

In France, President Sarkozy introduced legislation allowing a so-called 'three strikes' procedure to be used against persistent file-sharers. In its original format, it proposed the creation of a separate rights enforcement agency (suggested as a way of bypassing France's strict data protection laws) which would be empowered with the right to disconnect 'offenders', potentially from all ISPs, for potentially considerable periods. This legislation, while passed initially by the French government, was rejected by the French Constitutional Court, on the grounds, essentially, that there had to be a requirement for 'due process' in the decision-making process (i.e. if someone is wrongly accused of being a persistent file-sharer, what rights does that individual have to express their case, perhaps before a judge)? The legislation, in its amended form (featuring fines administered by a 'judge'), is, at the time of writing, back before the Constitutional Court for approval.

In the UK, the Labour government in 2008 encouraged the rights holders (e.g. the major record companies) to enter into a Memorandum of Understanding (MOU) with the government and the ISPs with a view to determining a policy on file-sharing. Absent agreement, the government indicated it would legislate on the matter. After publication of the Digital Britain report in 2009, and after initially backtracking on legislation, the government, with a new agenda from Lord Mandelson, is now pushing for a disconnection procedure to be adopted in the UK for persistent offenders.

Such a proposal is not, however, without its critics. The Featured Artist Coalition, a new lobbying organization set up by established artists such as Robbie Williams, Annie Lennox and Radiohead, among others, has instead called for so-called 'technical measures', such as bandwidth squeezing, to be implemented to frustrate potential file-sharers, although 'bandwidth squeezing' in itself also raises a host of other potential problems, including possible EU privacy law concerns and implications for end users' security and privacy (which may even contravene existing UK law such as the Regulation of Investigatory Powers Act 2000).

In the UK it is also possible that a s.97a Copyright Act 'authorization' case could be brought against ISPs to enforce rights holders' rights. A similar case was brought recently in Ireland under Article 8(3) of the Copyright Directive against Eircom (Ireland's biggest ISP); the matter was settled during trial between the parties, thus not providing any new useful legal precedent. A case is currently also being prosecuted in Australia by the Federation Against Copyright Theft against an ISP for authorization which, although it may provide additional fuel to the fire, may not necessarily set any new binding precedent. No one is quite prepared to put their head above the parapet for fear of being shot down!

Other voices, including this writer and consumer groups such as Consumer Focus (a statutory organization, formerly the National Consumer Council) think that remedies such as disconnection and/or bandwidth squeezing or other technical measures, certainly when applied without due legal process, are a leap in potentially the wrong direction. Penalizing the users of 'our' products seems a very counterintuitive way to build 'our businesses', particularly when one of the main criticisms leveled at the industry (and not unfairly) is that the industry was very late coming to the digital marketplace.

It would seem there are far too many 'holes in the dike' to plug; perhaps instead the industry should be metering 'the streams'? We already have the organizations and mechanisms in place. They are called the collecting societies. Fairly licensed and comprehensive digital services, under the auspices of perhaps a revised and invigorated (and more approachable) Copyright Tribunal, might just be the salvation this industry needs. As Radiohead has shown us all, there is still a great demand for well-packaged, well-marketed physical products (the £40+ box set), even when the digital rights are potentially offered for free!

The role of the music publisher

Sheet music used to be the primary source of income for composers, before the invention of the phonograph, and publishers were in the driving seat.

The introduction of radio and television brought with it a growing power base. Record labels realized the exposure this could bring to their artists. Of course, publishers were fast to realize the dawn of an international marketplace. In the days of Tin Pan Alley, music publishers supplied hit after hit to the industry. Great songs were often being recorded by several artists at the same time. Global marketing strategies didn't really exist for artists at this time, their local territory being big enough for any record company to break, but that was to change with the onset of radio and television. Often an artist in the UK would have a hit with a song, and then you would also see a different artist using the same song in the USA; for example, Dionne Warwick and Cilla Black, with 'Anyone Who Had A Heart', by Burt Bacharach.

The power base of the music publisher, however, slowly diminished as the power and strength of the record industry grew. In addition, as artists became arbiters of their own music it became harder for publishers to get artists to record songs written by dedicated writers and lyricists. Various techniques have been developed by publishers to assist in their involvement of song development, and you can see this occurring today. Publishers stopped signing writers (or certainly in the main) who had no way of helping to propel themselves into the industry. Publishers started signing more self-contained writer/artists and writer/producers (and this is still true to this day). Music publishers have looked for ready-made outlets for the song, and signing a writer/producer provided a ten-fold increase in the likelihood of a song being used by an artist, because of the producer's influence on the artist.

The current power base in the industry is changing yet again, and this has been evident for a while. As record labels have been driven together to survive by way of mergers and acquisitions to make sense of reducing income streams, they have also sought to change their business model and that has impacted on the role of A&R and label operations. If you like, the bigger the beast the more food it needs! Publishers and managers have had to step into the gap of A&R development in a greater way than ever before. This has allowed publishers to secure writers/artists at a more reasonable rate and to help guide and direct their development. However, managers have pulled away from doing this as it is just not a sensible business model for them. This business space is increasingly occupied by the music publisher and artists themselves who are eager to take their careers as far as they can. The independent publisher has been more active in this regard than the majors. I introduced one such

The Art of Music Publishing. DOI: 10.1016/B978-0-240-52235-7.10004-1

new artist, Newton Faulkner, to Peermusic through my A&R work at the Academy of Contemporary Music, at a very early stage in his development when Newton was a student at the Academy where I lectured. Peermusic's A&R man, Richard Holley, had independently gravitated to Newton as well and after several months of close development work Peermusic moved into a deal with the artist/writer. Two years later, their commitment, vision and investment paid off. 'Hand Built By Robots' reached No. 1 in the UK album chart (double platinum) and has received substantial success across Europe, Australia and the USA.

In doing this project Peermusic had crossed over the line of the traditional publishing role and into the world of the record label. They not only signed Newton as a writer, but also became his production company. Why? Simply put, they believed in him; the market was not ready for Newton, and no record label was prepared to put the time and effort into his development. The major labels do not have the time, or sometimes the expertise to dedicate to this role any more. I am rightfully nervous about saying this as a cover-all comment as there are some great A&R people out there, Colin Barlow for instance, but they are quickly disappearing. They came from an era when a label's role was more about discovery and development than anything else. Major labels now need a greater amount of market-ready material to feed the machinery that they have built. By doing so they are putting enormous pressure on each artist and every signing.

Peermusic's A&R (Richard Holley) and MD Nigel Elderton were convinced Newton had something special, and rightly so. In order for Peermusic to provide the recording and development support (which would formerly have come from a record label) they geared up their financial commitments and got on with it. Through their pluck and determination they proved that there is indeed a host of undiscovered talent out there. Could the power base in the music industry be switching back towards the music publisher? 'Rights management' is the key. You now have BMG Rights Management and Guy Hands referring to EMI as a 'rights management' company.

When the definition of a major label is about being able to achieve worldwide distribution and a global market, it's easy to see that there are some mighty competitors from outside the music industry only too willing to flex their muscles: companies that have worldwide distribution, better consumer profiling need content. Either the music industry will supply it to them, or they will develop other solutions, don't you think? The industry has already made one catastrophic mistake by ignoring the internet and what consumers wanted. Are they about to do it now on a business-to-business (B2B) basis and force media companies to walk into the space currently occupied by major record labels? Where will Apple be in five years' time? What about MySpace Records, EA Games Record division, and what of Nokia, Orange, Live Nation and other competitors? If worldwide distribution isn't a unique selling point (USP) of major labels then what are they? Leaders in creative development perhaps? No, that doesn't appear to be the case — they want more market-ready product. So it's not distribution or creativity. Then what exactly is a major label's USP? It's clear to see what they used to be but not so clear moving forward.

The other significant change taking place is in the contractual commitment between label and artist.

As a survival tactic labels have talked about and have introduced many 360-degree contracts. In essence they are looking to secure the rights of all income streams that an artist may develop (not just record sales). These include touring, merchandising, image rights and possibly publishing. We have yet to see how this might roll out and how quickly a case could be brought for 'restraint of trade'. I suspect it will all depend on how flexible contracts are when they go wrong. How can a lawyer value an artist's future touring and merchandising rights when an artist is just starting out? What level of advance and what percentage will be deemed fair?

The merger of the creative industries is another huge obstacle for major labels. They are not used to having other companies suggest or dictate terms. This shift in power initially came with the introduction of Napster, which after its launch in 1999 had over 50 million users illegally exchanging up to 1.3 billion files per month. Although Napster was shut down, its example encouraged other companies to find new means of distribution and new business models. In my research at that time it was as clear as daylight that the consumer wants 24/7 music availability, digital downloads, ease and speed of the music experience. The industry took no notice. In fact, my own primary research at that time highlighted the complacency with which leading music industry figures viewed the issue. There were letters from MDs saying 'it wasn't a publishing matter but something for labels to sort out'. The industry reaction was to sue the consumer (the customer) because of the negligence of the industry. Were music publishers using 'their best endeavors' to protect copyrights? No. Was the industry complacent? Sadly, yes.

Data transfer speeds and the data capacity of the microchip have led to, and continue to present, new opportunities for technology companies to influence music distribution. In fact, distributors of music can be regarded as direct competitors to the music industry as they require content to fulfill their strategic business developments, and the industry has again lagged behind in seizing this new potential alliance and income stream.

By mid-2007, the Global System for Mobile Communications (GSM) had 'served' calls, music and other data to 2.5 billion people, across 218 countries and territories,[1] and by 2009 this had increased to 3.5 billion.[2]

The industry's reluctance to embrace these opportunities cost it dearly, and handed the initiative to technology companies such as Apple. Apple has become the market leader in hand-held music technology and in the sale of online content. Since the launch of the iTunes Music Store in 2003 Apple has sold over three billion downloads. Apple continues to push the boundaries between technology and music ever closer with the introduction of the iPhone and iPod Touch. It exerts an influence

---

[1]http://www.dilanchian.com.au/ip-tech-e-biz/digital-music-technology-and-copyright-timeline.html
[2]http://www.zdnet.com/blog/btl/code-that-encrypts-worlds-gsm-mobile-phone-calls-is-cracked/
28942

over the music industry in the dictation of commercial terms, as recent experience in the industry's dispute over publishing revenue demonstrates.

Apple threatened to shut down the iTunes music store if an obscure three-person board appointed by the Librarian of Congress increased the royalties paid to publishers and songwriters by six cents per song[3] and further.[4] Do we really want a key distributor of music, Apple, to dictate royalty terms to the music industry? Their power base is very uncomfortable to the industry for exactly this reason.

Apple's dominance is being attacked by Nokia and Vodafone. Nokia launched a new initiative called 'Comes With Music'. The consumer is provided with unlimited music downloads for one year after purchasing a Nokia handset. Vodafone has continued its collaboration with MySpace by launching the 'Vodafone Music Reporter', an online profile set up to build a community based around Vodafone's 'Music Unlimited' festivals. Vodafone also played a key role in the release of Madonna's album 'Hard Candy'. Vodafone customers were given exclusive access to album tracks before the release date, as well as access to mobile content, ringtones, SMS tones and ringback tones. More recently, the industry has seen moves by O2 and Sony BMG to launch the first mobile operator/major label music store, while the Ministry of Sound has launched branded handsets which are preloaded with music, videos and wallpapers.

The potential success of these new developments is prevalent in the Japanese music industry. Japan has the only music market in the world where the growth in digital sales is compensating for a fall in CD sales. This is due to the vast array of music-related products for mobiles, accounting for 90% of digital music revenue.

The economic misfortunes of the world economy will undoubtedly play a more leading role in the influence of technology as retail collapses with the ever-burdening pressures of overheads and greater fixed costs compared with those of its internet rivals. I have no doubt that 2010 has witnessed more of a change in emphasis on how music is sold than at any other time in the industry's history. This is not all doom and gloom. It is change at unprecedented speeds and the winners will be those who can react in a more entrepreneurial way, with vision and business agility. Thankfully, the music industry is quickly coming to its senses as artists are ready and willing to jump ship. In 2007 Madonna and concert promoter Live Nation Inc. signed an all-encompassing deal after Warner Music Group Corp refused to match the Live Nation deal. Madonna said that she was drawn to the deal with Live Nation because of the changes the music business had undergone in recent years. Madonna went on to say that for the first time in her career, the way that her music could reach her fans was unlimited and that with this new partnership, the possibilities were endless. The deal with Live Nation encompasses future music and music-related businesses, including the Madonna brand, albums, touring, merchandising, fan club and website, DVDs, music-related television and film projects, and associated

---

[3]http://courtlistener.com/cadc/Recording-Industry-Association-v.-Librarian-of-Congress/
[4]http://www.loc.gov/crb/fedreg/2009/74fr4510.pdf

sponsorship agreements. Madonna is one of a succession of artists no longer choosing to sign to a major label.[5]

Prince, having turned his back on the internet, said that there was no money to be made and instead did a deal with a newspaper. The artist can make some good money, mechanicals are paid, newspaper circulation increases and all the middle men are cut out. Artists can see value in brand association and alternative routes to market where they can stay in control of their music and products. Only time will tell if the 360 model will work. Independent labels have used a version of this throughout their development but always provided greater flexibility, and it did not cover all aspects of income.

Music publishers have been able to spot new opportunities much more quickly than record labels and have looked for ways to do business with the media industries. Many publishers have tried to get writers/artists to re-record their hits once they are out of their re-recording restrictions in their artist contracts. Publishers are used to licensing rights; it's a normal part of business, and so when both the master and the music copyright are available this is a very powerful combination.

HOW DOES THIS HELP, EXACTLY?

Instead of the games company or advertising agency having to negotiate with and obtain clearance to use the recording from the original record label and separately from the music publisher, in this instance it can go to a music publisher, who may now own a new recording of the master and who also controls the publishing. Companies exist whose strategy is to re-record great hit tracks with 'sound-a-likes', and many of these alternative masters are licensed for use on a cheaper basis than the original products. This can be a very effective business model. It means that publishers have other allies in the marketplace to get their songs used if record labels are proving difficult to negotiate with. From 2000 onwards many such entrepreneurial companies were born and offered such a service. In addition, because music publishers are fully capable of licensing rights, publishers have been looking to make an agreement with labels, verbally or otherwise, to represent both sides of the copyright and, in doing so, offering what is known as a 'one-stop shop' solution. Chapter 7 on music synchronization will cover this in greater detail.

Record labels are really up against it at present. Their recorded masters are coming out of copyright after 50 years, which is a complete erosion of their asset base and the marketplace value of their company (look at EMI). Add this to the fact that too much product is already freely available on the internet (once it's out there you can't recall it). The only light at the end of the tunnel is some uniformed accountability being imposed on internet service providers (ISPs). There could be a long wait on this one, despite both the UK and US governments' apparent willingness to impose some sanctions.

---

[5]Source: http://today.msnbc.msn.com

Online copyright infringement will cost the UK music sector an estimated £200m in 2009, with some 7.3 million people engaged in unlawful file-sharing. Between the years 2007 and 2012 — according to research conducted by Jupiter Research — the cumulative cost to music companies will be £1.2bn. Losses of this order are clearly unsustainable, but the music community cannot tackle the issue alone. We need the support of internet service providers and the government.

BPI/Juniper, 2009[6]

While every lost sale of a recording is also lost mechanical income to the publisher, at least every new recorded version of a song is a new income stream, for the publisher and composer. Music publishing will therefore always better weather the impact of music piracy.

I would like to add here the latest news on MGM films. They have announced that all work has halted on the latest 'Bond' movie. This was indeed a great shock and we have yet to see the detail as to what has happened. However, one has to assume that the effects of piracy and diminishing returns are having an effect. It is something that affects all creative industries and all governments. The creative industries pay vast sums in tax revenues, from the individual to the corporation. If these industries continue to be damaged, government revenues will also be affected, and this is likely to be the tipping-point for meaningful discussions and concrete proposals. A combined creative strategic task force must surely be the most effective starting point, instead of fragmented discussions. We are all affected, but can all gain from an informative and collaborative internet e-commerce economy that empowers consumers but protects rights.

An update: in August 2010 Spyglass Entertainment indicated a rescue plan for MGM to take over management of Metro-Goldwyn-Mayer. Keep up to date with the story as it unfolds.

HOW A GREAT SONG SURVIVES

Two examples of this are 'Gimme Gimme Gimme' by Abba and 'Over The Rainbow', written by Harold Arlen and E.Y. Harburg.

Abba originally released 'Gimme Gimme Gimme' in 1979, with the record reaching the Top 10 in 11 countries. Since then it has been covered by numerous artists, including Erasure, Synergy, Yngwie Malmsteen and ATeens. It was also sampled by Madonna on her song 'Hung Up', and was included in *Mama Mia!* the musical and the film.

'Over The Rainbow' was originally recorded by Judy Garland in 1939, but has since been covered by artists such as Eric Clapton, Norah Jones, Louis Armstrong, Ray Charles and Eva Cassidy (to name but a few). It has been used on the X-Factor and American Idol on numerous occasions, and was performed by Kylie Minogue on her Showgirl tour. The Hawaiian Rainbow Singers and The Blanks have also

[6]http://www.bpi.co.uk/our-work/policy-and-lobbying/article/second-article.aspx

recorded ukulele versions of the songs, which have gone on to be used in the film *50 First Dates* and the television comedy *Scrubs*.

So, before I get into the functionality of the publisher's role, it is important to appreciate the current landscape, as this will affect how you work within it and what new initiatives you may bring to the table. Music publishing is an exciting and creative role, but it is also a business. In Chapter 14, we'll look at strategic planning and tools to assist in business development.

WHAT DOES A MUSIC PUBLISHER DO?

Today, music publishers are concerned with administering copyrights, licensing songs to record companies and others, and collecting royalties on behalf of the songwriter. Some of the more important music publishing activities are listed below.

Administration and registration of copyright

Irrespective of the type of publishing deal you sign, the basic function of administering and registering the copyrights the writer entrusts to the publisher is the most basic level of work that must be carried out. If this is not done then royalties are not collected, may go astray, or are accounted incorrectly to other people. I have known writers signed to major publishers where the publisher has failed to register a title and gone on and had a hit with it, only to discover to their cost that they had not carried out their basic duty. Needless to say the writer was pretty unhappy.

Mechanical royalties

The term 'mechanical royalties' initially referred to royalties paid whenever a song was reproduced by a mechanical device — pianos rolls, music boxes and then phonograph records (the old plastic ones). Some of you may never have even held one. The term now refers to CDs, audio cassettes, digital downloads, some interactive computer games, musical greeting cards, singing sunflower pots and the like. The amount of money a company must pay for a mechanical licence is generally set by the Copyright Royalty Tribunal. This rate is sometimes referred to as a 'statutory' rate. By way of example, the mechanical rate on CDs at this time is 8.5% of the published price to dealer (PPD). This is paid by the record label to the Mechanical Copyright Protection Society (MCPS; represented by PRS For Music) in the UK (an equivalent society exists in each territory). In the UK a mechanical fee is charged on digital downloads at a rate of 8% of the gross selling price less VAT.

Synchronization income

One of the largest growth areas in the industry is with regard to synchronization. The development of so many digital channels and different government mandates regarding outsourced content by television companies to smaller independent units has provided for an increase across all financial levels in the industry. By this I mean that a well-known song and an unknown song have an equal demand. Where budgets are small, companies defer to lesser known titles or new material, giving new composers and artists a chance to be heard via such media. Whenever a song is used with a visual image, it is necessary to obtain a synchronization licence permitting the use of that song. Music publishers will negotiate and issue synchronization licences to television advertisers, motion picture companies, video manufacturers, mobile communications companies, computer games companies, and so on. The resultant income will be shared between the writer and the publisher based on the terms agreed within the publisher/writer contract.

Print licence

Although sheet music sales have diminished over the years, many songs are still available in print form. You've probably got some at home: a guitar book, or a Beatles' songbook. The music publisher issues a print licence and collects this income from the sheet music company. The songwriter and lyricist receive a small royalty derived from the sale of the sheet music or a pro rata amount if it's one item in a portfolio of works. One of the best known companies in the UK is Music Sales.

Public performance royalties

A copyright owner also has the exclusive right to authorize the public performance of that work. This is why radio and television broadcasters must enter into agreements with performance rights organizations such as PRS in the UK and BMI, ASCAP and SESAC in the USA. These performance right organizations collect and account the income on behalf of songwriters and music publishers whenever a song is publicly broadcast. In the UK, 50% of the performance income always gets paid directly to the writer and if the contract with the publisher permits further income then the publisher also remits some of its income to the writer. An example is provided in Chapter 5.

Song and writer promotion

Often this job is carried out by the publisher or the writer's manager or agent — it involves convincing popular artists to cover existing songs, or perhaps convincing the BBC to use your latest tune in their next television drama. In the Introduction I mentioned Mike Collier and Freddie Faber. Both had an extraordinary knowledge of catalogues and were engaged by a variety of publishers to work their back catalogue. The trick was to keep getting covers and find new outlets for their copyrights; in short, to develop new income streams from existing copyrights.

Translations

Publishers may authorize translations to generate income from cover versions of a particular song in foreign countries. This is an area that is often overlooked. If the song allows (style), consider getting a Japanese vocal version done. Italian song-writers and labels have been very successful in Japan, where up-tempo music still works, and a great ballad with strong structures also works well across Japan and south-east Asia. These markets can provide extraordinary income. Ensure that the authorized translation is fee based and no allocation of copyright is given on that local version.

In summary, a publisher should:

- register and administer the copyrights for the territory provided under contract
- secure and protect the copyrights where possible and take such necessary action as required
- act to generate income (under the mandate of the contract with the writer)
- ensure that moral rights are upheld, as provided for in the contract
- collect and account royalties to the writer for the territory within the contract
- issue licences for the use of music to generate income
- market and promote the song/music and the writer — especially important under an exclusive writer contract. While specifics will not be in the contract it is expected by the writer and is an obligation on the part of the publisher. It must be fully discussed between the parties and generally forms part of the reason as to why the writer wishes to sign to the publisher. If an exclusive writer contract fails to meet expectations the writer will usually cite the lack of creative input on the part of the publisher. The publisher will often cite the lack of success in generating income with the songs provided or that the writer/artist's career failed.

I believe that publishers have a duty to engage with and be involved in satisfying the needs of new industry customers — the media industries. To avoid or ignore this duty, or to delegate this obligation, can only harm the writer and ultimately the industry. If rights are assigned to the publisher there has to be an obligation to protect these rights but also to develop new income streams in a proactive manner. Personally, I feel that music publishers have not been as responsive to opportunity and change as they could have been, again encouraging media companies wanting to do business.

Choosing the right publisher will depend on the point you are at in your career, and what your strategic objectives are (is it all about the money, or do you need/want career input and assistance with your own creative development?). Obtaining a contract with a music publisher and whether your royalties are paid on an 'at-source' or 'net-receipts' basis will in some part depend on the answers to the above. (At-source royalties and net receipts will be covered in Chapter 6.) There is no right way or wrong way; each decision will have its pros and cons, but ultimately it boils down to the people involved. In very simplistic terms, at-source royalties tend to be paid by major publishers to writers and net receipts tend to be paid by independent publishers. There will also be structures that combine both. There are differences in the characteristics and traits of majors and independents and these elements are just as important to consider. I've worked with many writers who no longer wanted to be represented by a major publisher, preferring the personal touch of an independent.

Then again, I've also known many writers who wanted major representation only, where the money was the key factor.

Point to consider

In my opinion, major record labels should consider what they are best at. It may well be that licensing of master rights to third parties is not one of them. Their publishing divisions, however, are eminently suited to this; it's a function they do standing on their heads. It makes sense to bring licensing under one roof, one department. Music publishers are more capable of dealing with such matters, and this would leave labels to focus on sales, distribution, branding and marketing. It may be that EMI is considering some similar strategic framework with their new appointment of Roger Faxon (currently head of EMI's music publishing division), who will now become chief executive of the whole company.

LEGAL COMMENT BY ANTHONY HALL
When public domain is no longer public domain

In my own practice I often work with producer clients who create 'new copyright works' out of existing works or adaptations of existing works. For example, when a music producer samples a well-known 'break' or musical/lyrical portion of a third party copyright work, say the hook from an established hit, adding new arrangements, perhaps additional chords, melody and lyrics, then provided there has been an element of 'originality' (i.e. a high degree of 'skill and labor' has been involved in the creation of the 'new work') then the producer will in fact create a new copyright work, albeit one based on the original work.

'Sampling' has become an established method for the creation of exciting new copy-rights; a prudent (and well-advised) client knows that he or she must clear both the copyright in the original recording and the underlying composition (or publishing copyright) forming part of the 'new work' before exploiting the new work. Dedicated agencies and lawyers have been involved in sample clearances for a number of years now and it is common for the owner of the original recording copyright sample to agree to receive royalties in the new recording and, to the extent such arrangements can be negotiated, the owner of the publishing copy-right in the sample may also agree a split of the publishing in the 'new publishing copyright', although not always, as demonstrated in the case of the Verve's 'Bittersweet Symphony', which sampled the Rolling Stones; in this instance litigation swiftly followed, courtesy of the infamous (and now late) Allen Klein. So my advice to clients is always to obtain the clear-ances, in writing and upfront, and to ensure the licences are well drafted and stand up to scrutiny. Failure to do so will allow the original copyright owners to restrain exploitation of the new copyright work, with potentially devastating consequences.

Public domain works are surely a different kettle of fish? After all, with public domain works, e.g. Beethoven's Fifth, the copyright has long expired and since passed into public domain, so anyone can presumably use such a work with impunity? Well yes and no, as the case of *Sawkins v Hyperion* [2005] EWCA Civ 565, clearly demonstrates.

In the Hyperion case the facts were as follows:

Dr Sawkins, a world authority on the works of Lalande (a composer of Sacred Music born in 1657, who died in 1726), spent approximately 300 hours working on 'performing editions' of four of Lalande's original works. Clearly, the original works of Lalande are now public domain and do not in themselves attract copyright protection. In 2002, Hyperion produced a CD containing recordings of the Sawkins' performing editions without the consent of Sawkins. The question for the court was whether the performing editions established a separate copyright or not. Sawkins acknowledged that he had not adapted or arranged or substantially modified in any way the original Lalande compositions, but had instead transcribed the music from the original scores into modern notation and interpreted the shorthand on the original scores for the benefit of the performers. Sawkins undertook this work by analyzing the original scores in libraries in Europe.

Referring to the judgment in *Walter v Lane* [1900] AC 539, where Lord Halsbury LC held that the law did not permit 'one man to make profit and appropriate to himself the labour, skill and capital of another', the court held that notwithstanding '... (a) Dr Sawkins worked on the scores of existing musical works composed by another person (Lalande); (b) Lalande's works are out of copyright; and (c) Dr Sawkins had no intention of adding any new notes of music of his own ...', the effort, skill and time Dr Sawkins spent in making the three performing editions were sufficient to satisfy the requirement that the performing editions should be 'original' works in the copyright sense, and thus, protectable. Hyperion lost, and appealed, though the decision was affirmed on appeal.

This sets an important precedent. First, copyright can exist in a new work based on a third party work, provided sufficient effort, time and skill has been involved in the creation of the new work. Second, such a copyright can exist even where the original work is public domain.

In the world of pop music, where 'mash-ups' of two existing works (or a new work and an existing work) are becoming more common, copyright may, following Hyperion, exist in the mash-up itself, even if the mash-up could not itself be exploited without the consent of the original owners of the original works. Going forwards, Hyperion may well assist producer clients in negotiating better terms for their part in the creation of the 'new work'.

Income streams

5

This chapter is clearly quite key. It's about how you will live as a songwriter or build your business as a music publisher. Remember these words forever: in any business cash flow is king. If you need any proof of this, look at the number of businesses that shut during the recession, and further, the number of business owners the banking sector failed to support that ran profitable businesses. Having a profitable business is only part of making your business successful. You have to make sure you have money in the bank; that it arrives when you expect it to. If you have a large debtor balance on paper and you look healthy and strong but in fact no one is paying you then your business is on a slippery slope downhill. Profitability and cash flow are two different parts of a business and both equally as important. When liquidity in the money markets is tight then cash flow becomes critical. Keeping costs under control and planning well ahead, having a strategy that also monitors external forces on a business is essential so you can plan how you may be affected by such changes.

I have personally assisted many companies as a mentor through this very difficult climate and in addition have been hands-on with my own family business beating the challenges the financial climate has forced on everyone. It is a constant battle in these precarious times and you cannot rest a moment when monitoring and predicting cash flow and sales.

In every respect the music industry is a business and probably one of the most difficult to work in. Many people on the outside see nothing but glamour and money, whereas in fact it's a very intensive environment to work in. However, the same principles of best business practice can be applied to every strand of the industry and then mixed in with creativity, passion, commitment, determination and all things entrepreneurial.

The music industry's money cycle is different in each sector of the industry. We're going to focus just on music publishing, but a good way to learn about money flow through the industry is to refer back to the universal PRS map in Chapter 1 and then attend key instructional sessions put on by a number of organizations discussed through the book. Being proactive in this way will help your knowledge base grow. Bit by bit you can then piece together the music industry.

So let's take a look at the route music royalties take and in addition the time royalties actually take to reach your bank account, as more often than not they can take a very long time. This is one of the key reasons that advances are given, between subpublishers and publishers and writers—they help to bridge that gap. Advances are just that, an advance against future income expected. The contract should always state that they are recoupable and not returnable.

The Art of Music Publishing. DOI: 10.1016/B978-0-240-52235-7.10005-3

We'll take a look at each key area of income and break it down. The music publisher has four key areas of income stream. Performance income is regarded as being the most steady and in the current climate has actually grown, supported by a very buoyant live industry and the licensing of many new performance delivery platforms. Performance income includes live and touring, radio and television broadcasts, and now expands into internet radio and mobile television and video, ringback and ringtone licences, podcasts and sites such as YouTube and Spotify. As technology drives new products so new licences will result and this is something that the industry needs to stay on top of. PRS for Music has recently joined forces with MySpace in a strategic relationship to promote the site as the online environment in which to share and enjoy new music. This is to help dissuade the use of peer-to-peer sites and encourage creativity and discovery in a 'safe' manner. Personally I think it's a great idea and could be part of the educational strategy needed to educate and safeguard copyright. The strategic framework announced by PRS for Music in April 2010 would look to:

> *Driving PRS for Music Membership* − *using the MySpace service to communicate to new writers and musicians the importance of PRS for Music membership.*

> *Benefits to PRS for Music Members* − *exclusive offers including songwriting master classes, competitions, and cross-promotional activities throughout the year.*

> *Driving MySpace Usage* − *demonstrating to existing PRS for Music members the value of using MySpace to engage both fans and the wider industry.*[1]

PERFORMANCE INCOME

PRS will collect all royalties generated from a performance of a song. What does this mean exactly? Simplistically, every time a song is played on radio or an artist performs a song on television, there is a fee payable by the broadcaster to the PRS and the PRS then checks its database to see who wrote the song and who publishes it. They then pay a share of these royalties (after first deducting their commission) to both the writer and the publisher. I'll explain a little further on a typical income stream.

Next time you go into a bar, a club or even a hair salon, you will see that these premises are all licensed to be able to play music from juke boxes or similar within their premises. Such premises operate under a blanket licence. How does this money reach the writer? The PRS will pay to publishers and writers a share of such money collected based on the pro rata use they determine by copyrights or publishers throughout an accounting period. It's not an exact science in that respect. This

---

[1] http://www.prsformusic.com/aboutus/press/latestpressreleases/Pages/
PRSforMusicannouncesstrategicpartnershipwithMySpace.aspx

includes music being played on radio, on television, in a shop, live performances or any public broadcast of the song. One-hundred per cent of these royalties are collected by the PRS through either blanket or individual licences.

- A blanket licence gives the user the ability to use specific rights contained within the song(s) or catalogue controlled by the society. The cost of the licence is affected by any limitations that are placed on the terms of the licence.
- Individual licences are granted for specific works; for example, if the work is to be included in a film. They are most often issued directly by the music publisher, who will want to negotiate the rate based on the specific use to which the song is being put.

Upon collecting the royalties the PRS makes deductions to cover the cost of administration. Table 5.1 outlines the deductions made on various broadcast income streams.[2]

For a full list of all deductions made by the PRS for Music or other societies such as ASCAP, BMI or SESAC, please go to their respective websites where such information is publicly available.

Having deducted an administration fee, PRS will then pay the remainder of the royalty to the publisher and writer: 50% is paid to the music publisher and 50% to the writer. If the writer is not a member of the PRS (or another eligible collection society), then PRS can be instructed to pay 100% of the royalties to the music publisher, who will then pay the writer directly. The 50% paid to the writer is not available to be used in the formula in recouping any advances. This is paid to the writer consistently, come what may, by the PRS.

What happens if you are on a 70/30 deal with the publisher? You are supposed to get more than 50%, so why is the music publisher being paid this? Very simply, the missing 20% forms part of the money that is paid to the publisher. This is done for two reasons:

- If the publisher has provided the writer with an advance then it is the percentage above 50% (that goes to the writer directly) that is used in part to recoup against the writer's account. In the contract this missing money is expressed in a way that is often at first confusing. It will state (using a 70/30 contract split as the example) that the publisher will pay to the writer 40% of the publisher's share. Forty per cent of 50% is the same as 20% of 100%. So the writer received 50% directly from the PRS, and the missing 20% (which is now referred to as 40% of the publisher's share) is accounted back to the writer if there is no advance to recoup, or once the writer is fully recouped. The writer will always end up with their contracted 70% at the end of the day.
- By law, and as a result of legal precedence set and ratified throughout the international collection societies, the writer will always get 50% of the performance income directly.

---

[2]These fees are correct as of July 2008.

Table 5.1 UK Administration Deduction Rates

	Rate (%)
Broadcasting	
BBC	12.5
ITV	14.0
Channel 4 and S4C	16.0
Channel 5	16.0
GMTV	16.0
BSkyB	14.0
MTV Europe	16.0
Other music channels	16.0
Other satellite and cable	16.0
Commercial radio	15.0
Public performance	
Popular concerts	20.0
– Up to a max. deduction per event of £1250.00	
– Concerts administered under the Live Concert Service at a per-act per-event fee of £125.00	
Classical concerts	20.0
– Up to a max. deduction per event of £1250.00	
Cinema	16.0
General live (non-concert)	20.0
Commercial discos and clubs	20.0
Other recorded	20.0
Public reception	20.0
Other	
Internet operators	12.0
Ringtones/ringbacks	12.5

Source: PRS for Music, current as at 2010 (please consult their website for up-to-date rates).

PRS pays its members on a quarterly basis, with the payment dates dependent on the form of broadcast and use (i.e. television, radio, etc.; see Figures 5.1–5.3).

In the case of blanket licences, the PRS uses sampling techniques to try to distribute the royalties evenly. Invariably this method means that not all writers and publishers will receive the royalties due to them.

Writers and publishers will receive a royalty every time music is played on radio (Table 5.2). The value of each play is dependent on the revenue received from and the amount of music played on each station. This is based on the reach and audience levels of each station.

Jan	Feb	Mar	Apr	May	Jun	July	Aug	Sep	Oct	Nov	Dec
July Distribution			Oct Distribution			Dec Distribution			April Distribution		

FIGURE 5.1

PRS payment dates for television (except for music television channels).

Jan	Feb	Mar	Apr	May	Jun	July	Aug	Sep	Oct	Nov	Dec
July Distribution			Oct Distribution			Dec Distribution			April Distribution		

FIGURE 5.2

PRS payment dates for radio, cinema, all live events except for concerts, recorded background music, music television channels, major network operator ringtones and Apple.

Jan	Feb	Mar	Apr	May	Jun	July	Aug	Sep	Oct	Nov	Dec
Oct Distribution			Dec Distribution			April Distribution			July Distribution		

FIGURE 5.3

PRS payment dates for all other companies licensed under the Joint Online Licence (JOL).

Again, let me just mention here that the rate of technological change is requiring that all such societies negotiate rates with new companies operating a variety of different business models. The following rates on streaming were approved for three years from July 2009.

For on demand streaming services, the headline royalty rate will increase from 8% to 10.5% of revenue in exchange for the per stream minimum being reduced from 0.22p (£0.0022) to 0.085p (£0.00085). Similar changes are also

Table 5.2 UK National Radio Rates	
Station	**Value Per Minute** [a]
BBC Radio 1	£17.68
BBC Radio 2	£19.35
BBC Radio 3	£14.74
BBC Radio 4	£23.34
BBC Five Live	£15.27
BBC World Service	£11.90
Classic FM	£4.04
Talk Sport	£0.61
Virgin Radio UK	£1.00
[a]Rates are subject to change.	

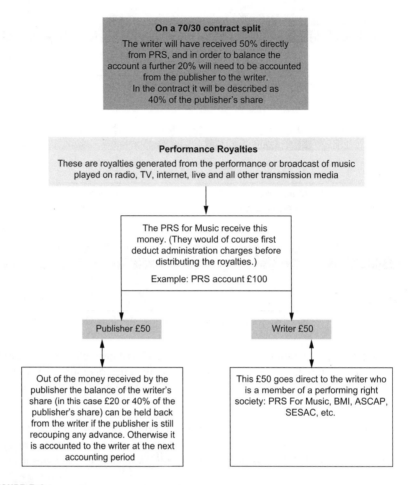

On a 70/30 contract split

The writer will have received 50% directly from PRS, and in order to balance the account a further 20% will need to be accounted from the publisher to the writer. In the contract it will be described as 40% of the publisher's share

Performance Royalties

These are royalties generated from the performance or broadcast of music played on radio, TV, internet, live and all other transmission media

The PRS for Music receive this money. (They would of course first deduct administration charges before distributing the royalties.)

Example: PRS account £100

Publisher £50

Writer £50

Out of the money received by the publisher the balance of the writer's share (in this case £20 or 40% of the publisher's share) can be held back from the writer if the publisher is still recouping any advance. Otherwise it is accounted to the writer at the next accounting period

This £50 goes direct to the writer who is a member of a performing right society: PRS For Music, BMI, ASCAP, SESAC, etc.

FIGURE 5.4

PRS for Music: royalty flow.

being made to the royalty rates and minima for Premium Interactive Webcasting Services and Pure Webcasting Services.[3]

The diagram in Figure 5.4 has been compiled on the assumption of £100 being generated by performance income, with the writer being on a performance royalty split of 70/30 with the publisher. For the ease of demonstration, no administration charge has been deducted.

Mechanical income was by far the greater income stream for many years, but internet piracy has seen this move into decline. It is easy to find industry statistics all

---

[3] www.prsformusic.com (press release).

over the internet showing this declining income stream. Of course, with declining revenues come declining taxes. This has been the main driver behind governments finally taking note. For governments, the maths is quite simple; now they are keen because they can see a direct correlation with lower taxes being paid to the treasury by the industry as a result of internet piracy.

The PRS for Music acts as an agent on behalf of its publisher and writer members, licensing four[4] of the six economic rights contained within their members' works. Essentially they are granting 'users' the mechanical right to make a copy of the song. They also grant users the right to manufacture and distribute the song for retail sale.

Record companies wishing to include songs on their audio products require a mechanical licence from PRS for Music. These are an AP1, an AP2 and an AP2A agreement. The financial status of the record company will largely determine the licence awarded.

An AP1 agreement is given to larger record labels with a proven financial history. They pay royalties due quarterly, based on the total number of records sold, less any returns.[5]

An AP2 agreement is given to smaller record labels. Payment is made on the number of records that are going to be manufactured, ensuring that its members receive payment before any records are pressed.

HOW IS INCOME GENERATED?

In simple terms, every time a record is sold in any format the record label has to pay a fee to the owners of the song (the songwriter and music publisher). This is equally true everywhere in the world. In the USA it's a little more complicated, with so many cents per track up to a maximum number of tracks, and then the fee can also be reduced further under a controlled composition clause.

This money is collected in the UK by PRS for Music and in the USA by Harry Fox. The money is paid over to the music publisher who, in turn, periodically accounts to the writer(s).

There is no initial fee due from a record company if they want to include a song on a recording, but they owe the writer (or whoever the writer has assigned their rights to) 8.5% of the published dealer price (PDP) or published price to dealer (PPD) of sales. The PPD is the unit price paid by the retailer when purchasing the product from either the record label or distributor.

For digital product the rights owner in the UK is due 8% of the retail price of sales. For music that is delivered passively, on a website, the rights owner is due 6.5% of the total revenues from a website.

---

[4]The right to make a copy, distribute, lend or hire, and the right to adapt.
[5]Returns are records that have been pressed, but not sold.

PRS for Music also offers a range of blanket and individual licences, including licences for synchronization use online.

You must make sure that you separate in your mind the artist who is signed to a record label and who may or may not be the songwriter. The artist will be paid a royalty by the record label from the sale of each record (this is totally separate). The record label has to also pay a fee per record to the copyright owners of the song.

I would like to add a comment here. Stop and think for a minute …. It makes me very angry when I see nerdy technologists or new media executives spouting such claptrap as 'music should be free — artists can earn their living from touring'! I hope all those idiots have come to realize that if you are the songwriter only and not also the artist a key area of your income is from record sales (mechanical income) and they are proposing that they forfeit that income forever. Maybe they would like to give our songwriters a piece of their pay check by way of compensation! No? I didn't think so.

Everyone has the right to be recognized for their work and paid for it, end of!

DEDUCTIONS

Upon collecting the royalties, PRS for Music makes deductions to cover the cost of administration. There is a wide range of deductions depending on the licence held with the user, but some of the key deductions are summarized in Table 5.3.

PRS for Music will then pay the remainder of the royalty to the publisher, who will pay it to the writer according to the terms of their contract. The MCPS pays its members biannually, quarterly or monthly, depending on the licence given to the

Table 5.3 Deductions made by MCPS

Licence Held by User	What This Allows	Deduction Made (Commission Rate)
AP1	The sale of records to the public	6.75%
AP2	The sale of records to the public	12.5%
AP2 A	(as above)	7.5%
BT1 & BT2	Blanket TV broadcasting agreement with ITVA and BBC	12.5%
BR1 & BR2	Blanket radio broadcasting agreement with BBC radio and independent radio programming	12.5%

user. Consult the collection society specific to your territory for local territorial rates.[6]

PRS FOR MUSIC — MECHANICAL INCOME STREAM FLOW

The diagram in Figure 5.5 has been compiled on the assumption of £100 being left over after the deduction of society commission rates. The writer is on a 70/30 contract with the publisher.

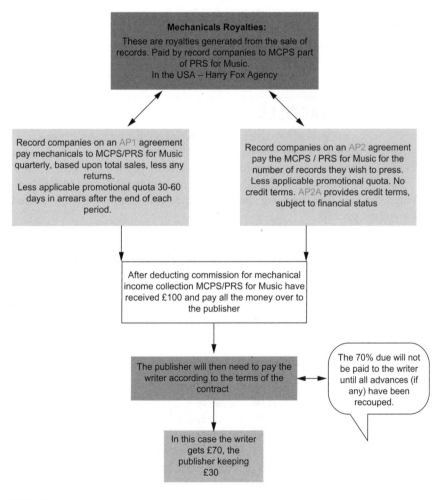

FIGURE 5.5

PRS for Music: mechanical income stream flow.

[6]www.prsformusic.com (mcps commission rates) — will vary, so check.

To clarify matters, in the case of mechanical income this is paid directly to the publisher whose job it is to account to the writer. If the writer's contract states that they are to be paid 70% of all mechanical income received by the publisher (70/30 deal) then at the following accounting period the publishers will pay 70% to the writer, having first deducted any advances paid to the writer from the writer's share of the money.

Once the writer has paid back the advance they will be accounted to in the normal way, receiving their full share (in this case 70%) of income from the publisher at accounting times specified in the contract. The amount due to the writer is specified in the contract. It will never be less than 50% and is subject to negotiation.

From day one the publisher receives its share of the money (as in the above example); any recoupable advances given (money paid to the songwriter) are paid back to the publisher *only* out of the songwriter's share.

ACCOUNTING PRACTICES

These can vary quite a lot these days but the standard industry accounting practice is twice per year within 90 days of 30 June and 31 December. Some publishers try to differentiate themselves in this regard. Kobalt Music in the UK is a good example and I hope more publishers will follow their example in time. Kobalt pays monthly and empowers its writers by being able to track their royalties, much like you would track a courier package.

> *Kobalt is a leading independent music publisher offering unparalleled online global administration of copyrights, pipeline advances and creative services to songwriters and publishers. Our unique, transparent systems were designed to better service the rights owner. Kobalt takes the guesswork out of complex global royalty collections and reporting, and enables clients to get more money more quickly, as well as access to flexible and accurate information.*[7]

SYNCHRONIZATION

Music publishers have seen a steady increase in synchronization income with the explosion of new media and how this is used and delivered to consumers. Such areas include television, film, computer games, television and cinema advertising, video and apps, and synchronization is one of the primary areas of income stream in today's media-driven society. As we have seen, a music publisher will collect performance and mechanical income through the collection societies, while synchronization income is usually dealt with directly by the music publisher, who will negotiate the fee and restrict the use to which the licensor wishes to apply the

---

[7]http://www.kobaltmusic.com/

music. The fee paid will be dependent on the length of time the music is heard, where it is placed in the film, program or game, and the time period – six months' use, five years' use or even 'in perpetuity'. The fee chargeable would increase if it was a key opening theme, and decrease if it was in the underscore. If they want to use a hit song in an advert, it will cost the advertising agency's client more money than if it were an unknown piece of music or song.

I should like to turn your attention back to cash flow for a minute. It is important to bring this up again. Royalty income streams can take 6–12 months to reach a songwriter from the local territory use of the music, and some 18–24 months, or longer, when considering international use. Because synchronization income is negotiated directly and the money is paid directly to the copyright owners it offers more immediate cash flow.

If you are starting a new publishing company, synchronization income can help to create immediate cash flow and must rank fairly high on your list of priorities. But, of course, you need songs or music to place. You will have read earlier about Platinum Sound, whose entire business growth was based on synchronization use. If this is now something you want to focus on, then your business strategy should make plans around how you achieve this: how you build up this particular network, who you need to know and how you communicate with them. Of key importance is finding out about synchronization projects that are coming up that you can work on.

Advertising agencies will be working on briefs and storyboards for their advertising clients; film companies will have movie scripts and will be engaging music supervisors to seek out and clear the music needed for use. Television companies will put music selection in the hands of the producers of the show or series, and also rely on the television companies' in-house music library (note that many television companies work under blanket licence agreements).

As I put pen to paper to complete this chapter, the latest figures are emerging on the performance of various companies, not least the Warner Music Group. I have highlighted part of their publishing figures as they provide food for thought. Are they an indicator of a future trend?

Music publishing revenue declined 1.5% year-on-year to $134 million, down 6.3% on a constant-currency basis. International music publishing revenue grew 3.8%, but fell 4.7% on a constant-currency basis. Domestic publishing revenue declined 8.6% from the prior-year quarter to $53 million.

Digital revenue from music publishing was up 85.7% from the prior-year quarter to $13 million, and rose 62.5% on a constant-currency basis. Mechanical revenue was down 2.4%, performance revenue declined 6.8% and synchronization revenue fell 4.0%.

On a constant-currency basis, mechanical revenue was down 9.1%, performance revenue declined 11.3% and synchronization revenue fell 7.7%.[8]

---

[8]http://www.wmg.com

Whatever industry you analyze, no one has been spared the effects of the current recession. Having said that, music and entertainment tend to fair quite well. What we see in the above figures is not just an adjustment due to internet piracy but a real effect of the recession impacting traditionally safe income streams. Synchronization income has been generally strong and increasing year on year, but here we see it moving downwards.

What could have caused this? This is likely due to television program commissioning being down (more repeats) and television companies wishing to pay less for music. The film industry is being impacted by both internet piracy and recession, and as a consequence may be pushing down the amount it is prepared to pay for synchronization use. It may also indicate that major film and television broadcasters are more willing to work with independents than ever before, as generally they will accept lower fees. There are some key points here that you can carry on investigating to see what trends may be emerging; if indeed this is a trend within a sector, or just Warner Music Group in isolation.

SHEET MUSIC

Most music publishers outsource their print needs to a third party company as this is a fairly small sector of the industry.

One such company is Music Sales, active in many music-related fields, which has offices worldwide. Principally, Music Sales owns, manages and exploits over 200,000 music copyrights. It is one of Europe's largest printed music publishers, distributing products worldwide from centers in the UK, the USA, Australia, Japan and Europe.

The source of Music Sales' copyright activities is the ownership of numerous international catalogues that include Campbell Connelly, The Sparta Florida Music Group, Bosworth & Co., Chester Music, Novello & Company, J. Curwen & Sons, G Schirmer, Edition Wilhelm Hansen and Unión Musical Ediciones (UME).

The group's Omnibus Press imprint is a prestigious world leader in books about music, covering everything from Grand Opera to contemporary pop music. Other print publishers include Faber Music Ltd and Oxford University Press.

In the USA the national publishers' association can help direct you to the key print publishers.[9]

ALEXIS GROWER – LEGAL PRECEDENCE

There are three important legal cases concerning music publishing that are of interest.

A Schroeder Music Publishing Co Ltd v Macaulay 1974
Tony Macaulay, at the time he entered into his contract with A Schroeder Music Publishing, was a young and unknown songwriter. The contract was a five-year contract which gave the

---

[9]http://www.nmpa.org/home/index.asp

publisher his copyrights for the whole world for the full length of copyright. When the total royalties earned by Tony Macaulay reached £5000.00 the agreement would automatically be extended for a further five years. Tony Macaulay's case was that the agreement was contrary to public policy as being in unreasonable restraint of trade and void. Schroeders contended that the doctrine of restraint of trade was inapplicable to their standard form agreement as contracts passed into, accepted and under normal currency of commercial relations did not require justification under a public policy test of reasonableness.

The court held that the restrictions in the agreement were not fair and were unreasonable in that the publishing company was not required to publish any of the compositions and the writer could not earn anything from his ability as a composer if they chose not to publish the compositions. The writer was the weaker person in the negotiations leading to the contract being signed. The House of Lords said that it was in the public interest that an individual should be free so far as practicable to earn a livelihood and to give to the public the fruits of his particular abilities. Nowadays in publishing contracts, if the publisher does not publish and exploits the writer's compositions then a writer can serve notice on the company and obtain reassignments of those compositions so the writer can publish them elsewhere. The case also dealt with receipt decreasing deals, whereby the publisher would subpublish with its own company in another part of the world to the detriment of the writer. For example, if publishing company A has a 50/50% deal with a writer and then passes on that publishing to its own affiliate in, say, France and does a 75/25% deal for that territory in its own company's favor, only 25% comes back to the publisher so that the writer gets 50% of 25%, i.e. 12.5% of the royalties, but the company picks up 87.5% of the royalties earned in the territory in question. Macaulay went on to write many hits, such as 'Build Me Up Buttercup' with Mike D'Abo.

The Stone Roses Case 1991

This case concerned The Stone Roses, who signed to Silvertone Records Ltd and Zomba Music Publishers Ltd in 1988. The case was heard in the High Court. The Stone Roses was a young rock group at the early stages of their career. The recording and publishing company is generally known as Zomba. Although The Stone Roses had the services of a lawyer available to them via their manager he was not a lawyer experienced in the music industry.

When Zomba wished to sign The Stone Roses they sent a highly experienced lawyer to Manchester to meet them. Negotiations took place but there were no changes sought by The Stone Roses in the commercial terms offered, although later on there was a request that the royalties in the publishing agreement be improved, which was turned down. The lawyer for Zomba was of the view that his suggestions for a deal were an 'opening position' and he could well have been persuaded to increase this if there had been any real pressure to do so or if The Stone Roses had the advantage of an experienced music lawyer. The fact that The Stone Roses and their people failed to seek to modify or improve any terms on offer shows the inexperience of those advising them. Indeed, Zomba, at that time, when they had heard agreements had been signed, was concerned that the lack of substantial negotiation could have made the contract unenforceable. The judge in the case stated that the services of a music lawyer in contracts of this nature are not the same as in many contracts, when a lawyer just explains to the client what the contract says and ensures, where appropriate, that the contract represents what the clients want. Within the music industry music lawyers go much further. They have the expertise to appreciate many of the terms, something of the state of the market, and the state of the law on restraint in entertainment contracts, which in recent years have been developing rapidly. Music lawyers habitually get involved in negotiations and know where it is right to and where they are able to put pressure on the other side so as to thrash out an agreement that is fair to both sides. The lawyer for Zomba knew that while he was expected to follow Zomba standard terms he would have been able to obtain authority to make concessions. Indeed, that lawyer was concerned that the absence of any serious negotiation might render the contract unenforceable. In his evidence the lawyer said that had there been competent representation for The Stone Roses there would probably have been negotiation of concessions on a number of important clauses. For that reason the judge held that the contract was not fair to The Stone Roses, was oppressive and imposed unjustified and unjustifiable restraint of trade, and set both the recording and publishing agreement aside.

The *Whiter Shade Of Pale* case

As recently as July 2009 the House of Lords considered the *Whiter Shade Of Pale* case. This was a case in which claimant Matt Fisher was seeking a declaration that he was the co-composer of the composition. The interesting fact is Matt Fisher waited 38 years to have the case dealt with by the court. Although a defense was raised that the repellent had not brought the case in good time and therefore was barred because of the lateness of the action, it was noted that Matt Fisher did not seek royalties going back after the six-year limitation period. What he wanted was to be registered as the co-composer. The evidence showed that the composition 'A Whiter Shade Of Pale' was composed in 1967, when Matt Fisher was only 20 years old, and at that time was considered in law to be an infant. The composition has been remarkably successful over the last 38 years and there have been 770 versions of the work performed by other groups during that time. Accordingly, Matt Fisher was awarded 40% of the song.

What is interesting with regard to this matter is that more recently, in Australia, a court has awarded the original composers of a song (a favorite with Girl Guides in Australia) in respect of a major hit by the band Men at Work. It was claimed that the composition 'Down Under' contained part of the track 'Kookaburra Sits In The Old Gum Tree', which was written by a teacher, Marianne Sinclair, for a Girl Guide jamboree in 1934. The claim succeeded and Men at Work and their publishers are having to pay substantial monies to the publishers of the original composition.

Introduction to contracts

This chapter will look at a broad array of contracts that are commonly used within the music publishing arena. Please note I am not a lawyer, but have operated within the world of business affairs for many years. I use lawyers in a variety of ways; sometimes to assist in concluding negotiations and always in the final drafting of the contract. This chapter is designed to help you to become more familiar with terminology and to take away the fear factor. It will help you to understand what is being asked of you, and areas that you could focus in on regarding negotiations. It's a great starting point and will make negotiations and discussions far more enjoyable. It is essential to understand the basics of what is being asked of you and their impact on your career as a songwriter. I would always advocate the importance of seeking proper legal advice when entering into any form of contract. It is also important to note that the lawyer that you choose to represent you must be practicing in the music industry. You cannot use a friend of the family whose area of expertise is in conveyancing (for instance), as they would not know what is commonly acceptable in the industry, and the areas of negotiation. I come across this regularly and it's just so stupid to risk damaging your career and your ability to earn a living by cutting corners in this way. This chapter will take you through key elements of the language and legal phrases used, what they mean and how they affect you, the writer, and I will also provide the perspective from the publisher's point of view. Remember, successful negotiations are when both parties step away from the deal feeling positive and happy with what has been agreed. What each party wants can be quite different. Therefore, it is useful to see and understand both sides of the coin.

There exists a broad array of contracts for all key areas of music representation and exploitation. Each carries with it specific structures depending on the use to which they are being put. As such, they can vary by way of complexity and length. Contracts could range from five pages to 30 pages. In general, the greater the perceived risk in the contract, the greater the complexity. For example, if Universal Music were issuing a five-year exclusive writer contract, with complicated advance structures and minimum commitments, then the complexity, length and risk factors would in some way be measured by the length of agreement because both sets of lawyers will be working to protect the interests of their respective client. No contract will ever be the same as the next. That is why it is so important to be represented by an experienced music industry lawyer. In addition, new areas of exploitation are developing regularly and how the industry is licensing these areas is subject to change. This has an impact on you. Each set of negotiations is relevant to the exact situation in which you find yourself. The successful outcome of any negotiation will

The Art of Music Publishing. DOI: 10.1016/B978-0-240-52235-7.10006-5

depend on the expertise of the lawyer and your own bargaining position and the added value that you may have brought to the negotiations.

This might include added-value areas such as:

- Artists want to cover (use) your songs which are signed to a label.
- You've co-written songs with established writers in the industry.
- Perhaps an established record producer wants to record and develop you.
- You have industry support in radio.
- You write your own music and labels are keen to sign you as an artist.
- Something you've written is being used in a television commercial, a film or a computer game.
- An experienced manager is now representing you.

The greater the added value the stronger your negotiations will be. Why is this? Simply, the publisher will be considering the risk factor of any deal (this is usually based on the amount of money you're trying to leverage) versus the potential return on investment that the publisher is likely to see. If they can see that others believe in you, and have invested their time and energy, then this starts to paint a picture about the direction of your career. In addition, and obviously key to this, is how good your music is. If you only have one good song in your repertoire it will be harder to convince a publisher of your longevity but it might be enough to provide a stepping stone into gaining their support to see if you grow and develop. If you were an artist this might be described as a development deal — as a writer, a publisher may invest in some time with you and put you in some co-writing sessions with other writers signed to them to see how you get on. This may result in their offering you a deal. Understand the perspective of the music publisher.

One of the reasons you need a good lawyer is that they are objective about the situation. It is for them at this point all about 'the business of music', getting the best deal for their client (you), and it's not about the notes on a page. A good lawyer will work to do their best for you as a new songwriter in the same way they would look to do their best for an established client. Clearly, when you bring greater 'added value' (success) to the table this is a stronger negotiating tool. While they have to stay focused on the job in hand, that's not to say that the music is not important. In my experience the best lawyers are also extremely committed and passionate about music and enjoy seeing their clients do well.

Your lawyer is there to help you get the best deal based on all the factors we are about to discuss. It is so important, however, that both writer and publisher understand what each is setting out to achieve. I will therefore flip from one to the other and try and represent both negotiating perspectives. Each contract is different because each negotiation will be different. The music publishing contract is there to assist and reward the writer but to also protect and provide a good return on investment for the publisher.

Ultimately, if the job has been done well, at some point all of the writer's copyrights will revert to the writer, and the publisher should have recouped the advances and made a good business return on the investment, the writer being in the black and earning royalties. If the relationship has been positive for both parties then the publisher will want to renegotiate with the writer to continue to represent that person and the writer will consider this strongly. It is human nature to consider that the grass is always greener elsewhere. Just remember, the music industry is built on relationships and if you have a good one and your contract is suitable there is no need to change. Remember too that an advance is an early payment on future expected royalties: it is not additional money and it has to be recouped from future income.

Despite all that you may hear, there are many very happy writers with great relationships with their publishers and they may choose to stay with their publishers for years beyond their initial contract. There are, however, many different types of publisher (majors and independents) with different cultures and ethos. It is important, therefore, that a writer finds the right team or individual to represent their work. The churn factor (staff leaving) within major publishing companies is far less than at record labels, but this has to be considered.

Before we start to look at each type of contract we should establish the basic language used in contracts. I'm sure you'll agree that once the language used has been demystified the reading of contracts becomes much more straightforward. For those who are totally unfamiliar with the layout and language of a contract it can be quite a daunting task, so this process will assist in calming those fears. There exists in all contracts a template of familiar language. The same or similar language is used across many forms of music industry contracts, and so as we go through music publishing contracts many words will also crop up in other contracts. Let's take a look at some of the most usual words or phrases.

CONTRACTUAL LANGUAGE
Contracting parties

All contracts will start with a clarification as to the contracting parties, i.e. who is entering into the contract. As an example, in a Single Song Assignment it will state the XYZ Publishers hereinafter known as the Publisher and Joe Bloggs hereinafter referred to as the Writer. Throughout the rest of the contract it then refers to the Publisher or the Writer.

Date

The date on which the contract will have been signed. This date will provide a fixed point in time against which all further references (within the contract) will be measured. For example:

This AGREEMENT made as of 1st December 2010.

Term

All contracts will specify a term. This is essentially the length of time for which the contracting parties may be signed to the contract. This is not to be confused with how long the contract may continue to represent all the song copyrights for. The term will commence from the date of the contract.

By way of example, after stating the contracting parties the contract will state the term, e.g. the term is for five years. It might then also state that the copyrights are retained (retention period) for five years after the end of the term: the term commences on the date of the contract. They are therefore now linked together. The overall length of time that the copyrights will be assigned (assigned or licensed) to the publisher will be the term plus the retention period, in this example $5 + 5 = 10$ years, but the writer can leave the contract after five years if the terms within the agreement have been met. The songs, however, remain with the publisher for the remaining retention period.

So now consider this. You've signed a deal with Universal Music and the contract term has now expired. You can now enter into a new deal with a different publishing company for any new material. However, the material written to date will remain with your previous publisher until the retention period has expired.

Why is this? The publisher has taken a risk in signing you and you will have received advances through the contract. It takes a publisher time:

- to receive income from around the world
- to have time to activate 'covers', and other uses of your songs
- to make a profit and a financial return.

At the end of the retention period the copyrights will return to the writer. So in the future you will be able to do a deal on your back catalogue as well as future writing. This is the structure you should always aim to be in. In the past, writers signed contracts (which were common at the time) on their songs for 'life of copyright', which ultimately means that those songs remain with that publisher for your life plus x years after the death of the composer (see Chapter 2).

Issues that may affect term and retention include:

- The minimum commitment not being met (the wording on minimum commitment can be very onerous and care must be taken to fully understand what is being asked of the writer).
- Covers the publisher has secured may provide an extension to the retention period for those copyrights.
- If the contract has not recouped within the term, an extension might be built into the contract tying the writer in until recouped; this will or should be capped up to a maximum period.

These three points get used a lot and it's so important that the mechanisms used by a publisher to lower its risk in having signed you are 'reasonable'. This again is where you need a good lawyer, but you also need to understand this yourself and its likely impact on you.

Ask yourself, if you were the publisher and not the writer how would you want to protect your investment? Is the publisher being reasonable or unreasonable?

Territory

The contract should state to which territory the contract is binding. For example, the contract will provide an assignment or a licence of your rights perhaps just for the UK market, European market or world market. The territory can be an individual country, a group of countries or the world.

For example:

'Territory' shall mean the world and all parts of the universe.

The universe features in most contracts due to satellite and any other transmission systems outside earth that may occur − or even a potential moon base!

Royalties

Royalties refer to how each of the income streams is split between the writer (copyright owner) and the publisher. The contract will show performance income, print, mechanicals and synchronization as separate areas of potential negotiations. Understand what these key areas cover. There may well also be a phrase that says 'and any other income', a general statement as an overall 'catch-all' for any other income streams and new uses that come into being. Royalties may increase or decrease in your favor in specific and defined situations in the contract. It is important to know when and why.

For example, the contract may say you have a 70/30 split on mechanical income (the writer being on 70%), but it may go on to say that on covers the publisher secures this may drop to 60/40 (the writer being on 60%). This is a mechanism that helps smaller independent publishers to provide a reward to their local subpublishers for obtaining local activity for them, and major publishers reward their other offices if they have helped to secure new income streams on non-locally signed writers to them. This is a fairly usual situation to face. Hey, but aren't you lucky if you have an active publisher getting you local covers across the world?

Advances

Any advances paid under the terms of the contract are usually only recoupable − make sure they are not returnable. This would only be enforced if money was procured by way of deception or fraud. There is a variety of ways the advances might be paid. Simply, however, there is usually an amount on signature (to validate the contract) and then perhaps at key intervals instigated by time (e.g. on the anniversary of the contract) or by way of a roll-over advance (an initial advance paid that subsequently is recouped will trigger another payment within certain parameters of the contract; note that this is not usually in the last 12−18 months of the contract term as publishers will not want to load their risk if not recouped close to the end of the term of the contract).

How advances are provided (and negotiated) is complex as they can also be tied into a number of other factors. These can include the number of songs delivered, the percentage of each song controlled, hits created (chart positions) and pipeline income. These elements can also have a complicated tie-in with the term and overall length of control within the contract. This is why it's essential to work with a lawyer who has current industry knowledge and experience. The two things you should never skimp on in your business life are a good lawyer and a solid accountant.

Rights

The rights within a contract are usually assigned or licensed.

Assignment

Assignment is the transfer of ownership of rights by an owner to someone else. You can assign all or some rights.

An assignor is the owner of the rights. The owner no longer owns rights when they are assigned.

The Writer hereby assigns to the Publisher.

Typically you'll see this wording in an exclusive writer contract or in a single song assignment.

Licensing

The copyright owner (songwriter) holds onto the rights and only gives the publisher permission to do certain things with the copyright. By way of example an administration contract is done under a licence and the publisher would be charged with registering, administering and collecting royalties in a given territory for a stated term (period of time). An administration contract is as it states, a contract to allow the administration of copyrights. Typically this is a business-to-business deal, i.e. between publishers and between territories.

But there is nothing to stop a writer who wanted to keep more control over his or her catalogue and was considering self-publishing (usually a more experienced writer, with a developed catalogue, but who didn't have the time to register songs and track royalties) deciding to engage a publisher instead. In return for the publisher's services, the publisher would be paid an administration fee, usually ranging from 10 to 15% of the gross income received by the publisher during the term of the agreement (note that current economic pressures may limit any deals where only 10% is charged). There are no advances usually paid and there is no 'creative' publishing involvement, i.e. the publisher generally would not 'work' the catalogue to create new income streams, but only collect upon the activity of the catalogue. If a more involved relationship was wanted then this would be done either under a subpublishing agreement where advances are usually paid, or by a songwriter signing directly to a publisher under an Exclusive Writer Agreement.

Accounting period

In the industry this is typically semiannually, i.e. 30 June and 31 December, within 90 days. However, there are plenty more frequent accounting practices appearing and, in my opinion, it is something that should be improved on throughout the industry by way of standard practice. Publishers such as Kobalt Music in the UK account monthly and provide a clear tracking system that their writers can use.

Going off subject for a moment — we've seen how over the past few years courier companies have empowered their customers to track parcels online. Imagine if this were brought into the music industry. I see no reason why the writer could not be given greater transparency and speedier payment structures as standard. I also see no reason why independent labels and publishers should continue to suffer from 'radio sampling' as a guide on which performance income is paid. Every play should be accounted and paid out. To do otherwise kills the independent label or publisher working hard to secure radio plays and build success. One of my students — a very bright young man — has come up with an idea that could solve this issue.

I also see no reason why publishing companies and societies couldn't develop such accounting and tracking transparency. This would provide real choice for a songwriter or catalogue owner when looking for the right company to represent them. This could be a powerful sales tool and unique selling point (USP) when describing such a service. One such company that has embraced this is, again, Kobalt Music.

Audit

The writer and a qualified accounting representative have the right to check and examine the books of the publisher to ensure that they have been accounted to correctly. The contract would provide an annual right to audit providing the publisher is given due notice. Carrying out an audit is a costly exercise and you are not likely to have a professional accountant undertake this task for you unless you feel there is a good reason. If you've had a massive hit, then it's prudent to arrange an audit at some point when royalties are flowing fully or if your accountant or lawyer feels you have not received sufficient payment for the exposure of your music. Most audit clauses would include a provision to cover the reasonable costs of an audit that you have actioned, should you discover any underaccounting. The amount of underaccounting that has to have taken place before such audit costs are paid will be specific to the contract. This amount is usually described as a specific figure and/or a percentage of royalties that have been underaccounted.

For example:

Any such audit shall take place during regular business hours at the location where such accounts and statements of the Publisher are maintained. Such audit may not be made more than once with respect to any particular statement. In the event that any such inspection reveals a discrepancy in favor of the Writer of £2000 or 5%, whichever is the higher, then the Publisher shall pay the Writer's reasonable inspection costs (excluding travel, subsistence and accommodation).

In the event that a deficit in excess of 10% of total monies in a particular period is showing to be outstanding the Publishers will pay to the Writer forthwith, together with all outstanding monies and interest thereon, the reasonable cost of such inspection provided that the cost of such inspection to be borne by the Publisher shall not exceed the total amount of such deficit.

Legal jurisdiction

All contracts will be bound under the legal jurisdiction of a given country and therefore the courts of that country. This clarifies which country you would have to fight a legal battle in if you had to defend or take action on a contract at court. Therefore, if you are British ensure that all contracts that you sign are under English law and under the jurisdiction of an English court. If you are based in America then contracts will come within the US legal system.

Summary

We have looked at such language as:

- Contracting parties — those to whom the contract relates.
- Date of contract — the date provides the start point to which all other time-frames relate.
- Term — the length of time you are bound by a contract.
- Territory — which countries the contract has control over and is applied to.
- Royalties — performance, mechanical, print, synchronization, other.
- Advances — legitimize a contract; they are recoupable (and should not be returnable).
- Assignment — a transfer of rights; an assignor is the person or company to whom these rights are transferred.
- Licence — permission granted to carry out certain functions, or the act of licensing.
- Legal jurisdiction — where (the location) any dispute would be actioned at court.
- Audit — your right to check the statements of the publisher.
- Accounting period — how often you will receive a statement and monies accounted to the writer or recouped against any advance provided.

All contracts should seek confirmation that the writer has had independent legal advice (this is relevant across all industry contracts). All publishers and record labels have to ensure this has happened, otherwise their contracts are not enforceable. Legal precedence exists on such matters.

These key terms will crop up in many contracts and so a familiarity of this language will help you to develop a good basic understanding and confidence in what is being discussed and how this affects you.

TYPES OF CONTRACT

There are several types of music publishing contract that you are likely to encounter. The key contracts are:

- single song assignment
- exclusive writer contract
- co-publishing agreement
- administration agreement
- subpublishing contract
- collection agreement
- synchronization agreement.

Single song assignment

This is a writer to publisher deal and is a simplistic contract, often referred to as a 'specific' agreement. It is used as a contractual mechanism to publish individual songs. It could be one song or several, but it is specific in nature. The composer is not signed exclusively through such a contract, but is free to enter into as many single song assignments as the writer wishes. Different contracts represent different song titles. The writer would generally assign his or her rights to the publisher, typically for a specific period, the 'term'; when the term has expired a collection period would follow.

Collection period

This is the period at the end of a contract to allow the publisher to receive monies accrued during the term of the contract, but which have not yet been accounted to the publisher. This provides a mechanism to accommodate the lengthy time it takes for royalties to flow via the record labels, television or radio stations and other outlets (in any country) to the local collection societies. The royalties then transit on to the local subpublisher or back to the original signing territory and then through the local performing and mechanical rights society (PRS for Music in the UK, ASCAP/BMI/SESAC and Harry Fox in the USA) on to the writer and publisher, based on the terms of the contract and society rules.

The publisher is charged within the contract to collect and administer the royalties from the songs, to protect the copyright on behalf of the writer and to use their best endeavors in creating new income streams. Advances, if any, are generally small for a single song assignment. A lot depends on the activity (if any) on the song(s) to date, or likely activity. The period of control should be fair and reasonable. If the financial risk to the publisher is little or none then the contract can be structured in such a way to be very fair to both parties. Such structures might include a reversion of the copyright to the copyright owner (writer) if no activity on the song is obtained by the publisher within a reasonable period. This may often be considered to be two or three years. If the publisher does get a cover on the song or other activity described within the contract then the term (or rights period) could be extended by a number of years, for example five or seven years, and even this could have an extension of time which could be tied to the earning capacity of the song during that period. If the song generated more than x to the writer then there could be a further extension to the term (rights period) for the publisher having done a good job, providing of course all other elements of the contract had been properly managed, i.e. accounting had been done on time and the publisher was not in breach of contract in any way.

Examples of wording relating to the contracting parties

> *THIS ASSIGNMENT is made the 15th of January 2010*
>
> *BETWEEN Bobby No Songs, herein after referred to as the Writer ('the Writer') of (address), AND Make Me Some Money Music of (Address), herein after referred to as the Publisher ('the Publishers'), of the other part.*

Example of a clause clarifying how long a title or specific titles are assigned to the publisher

This is a typical description. It only relates to specific titles and not to the writer. The writer is not signed exclusively but has agreed to let a publisher handle a selection of songs, in this example for 10 years from the date of the contract. The publisher recognizes that the rights hereby assigned are subject to the rules and rights of PRS and the writer's affiliate societies, assuming he or she is a member of PRS. All writers should look to be members of their local society (examples listed earlier in the book). If the writer is not a member then the publisher will collect 100% of all performing income instead of the writer being able to receive 50% directly through their local society and worldwide affiliates.

> *In consideration of the royalties payable to the Writer hereunder the Writer hereby assigns to the Publishers the copyright and all other rights now or hereafter known or recognized under any law, other governmental regulation or judicial decision (subject as hereinafter provided) in the musical work(s) listed in the Schedule hereto including the title(s), words and music thereof and all other rights of whatever nature (whether now known or hereafter devised) including without limitation any and all rental and lending rights and cable retransmission rights ('the Works') for a period of (10) ten years from the date of this contract ('the Term') throughout the world and the universe absolutely free from any adverse claims by any third parties PROVIDED that if the Writer is a member of the Performing Right Society Limited (PRS) or other affiliated society the Rights hereby assigned are subject to the Rights of PRS in the performing Right in the Works arising by virtue of the Writers membership of PRS or otherwise but include the reversionary interest of the Writer in such Rights of PRS expectant upon the determination by any means of such rights subject to the payment to the Writer by the Publishers of the Writer's share of all fees received by the Publishers after such determination in respect of such*

rights, such share to be the same as the share previously payable to the Writer by PRS in respect thereof.

Example of an accounting clause

The Publishers will within 90 days of 30th June and 31st December in each year calculate the amount owing by them to the Writer hereunder at each such date and will pay such amount to the Writer after deducting any unrecouped advances (or other chargeable costs) made by the Publishers to the Writer and any sums owing to the Publishers by the Writer. All sums payable shall be subject to the deduction or withholding of all taxes required to be deducted or withheld under the laws of any country or territory and exchange control regulations of any country or territory from which those payments emanate.

This type of contract would conclude with a list of terms clarifying that it's governed by the laws of England, and a statement of key points as to how the contract should be operated. If you are based in the USA then it would come under the jurisdiction of the applicable state and say that in the contract.

(a) *This Agreement shall be governed by and construed under the laws of England whose courts shall have exclusive jurisdiction.*

(b) *If any part of this Agreement shall be invalid or unenforceable it shall not affect the validity of the balance of this Agreement.*

(c) *The waiver by the Publishers of any breach of this Agreement shall not in any way be construed as a waiver by the Publishers of any subsequent breach whether similar or not of this Agreement by the Writer.*

(d) *Nothing contained in this Agreement shall be deemed to create a partnership between the Writer and the Publishers.*

(e) *The Agreement contains all of the terms agreed between the parties at the date hereof. Any variation or amendment to this Agreement shall only be effective if agreed in writing and signed by or on behalf of both parties.*

The contract should allow for any copyrights assigned to revert to the copyright owner should the publisher go into liquidation or appoint an administrator and such appointment not secured within 30 days. It is very important that a reversion of rights exists; otherwise, the liquidator could sell the copyrights to another publisher and your contract would be in the hands of another publisher, which may not be what you want.

Summary

A single song assignment is a contract that would be used to represent single songs or a collection of songs, but it does not contract the writer to the publisher in any further way. It is a great contract to be used when you're first starting in publishing and selectively picking up songs that you want to work. There is little financial outlay for the publisher and the risk to both the writer and publisher can be minimized by reference to the above points.

Exclusive writer contract

This is a far more serious contract than a single song assignment.

This contract will usually come with a health warning at the top (I jest). It will state that this is a life-changing contract and you must take advice from a lawyer (true). This is to help prevent any such claim you may have in the future as to 'I didn't understand what I was signing'. Again, I must emphasize: please don't sign a contract without having a music industry lawyer represent you!

What happens if you have no money for a lawyer? This puts the other party (the publisher) at an unfair advantage and so it is not unreasonable for your lawyer to request a contribution or budget for legal fees from the publisher (this can also occur between artists and labels).

This type of contract is a writer to publisher contract. The nature of this contract is that the entire output of a writer falls within the body of this contract. There will be serious undertakings by both parties, but particularly by the writer, especially if the contract includes fairly weighty advances. Any writer should understand that the only way the publisher can recoup its investment in the writer is by recouping against royalties earned from the activity of the songs/music. The advances are not returnable, only recoupable. This type of contract is on a par with an artist signing a record contract. Both can be life changing in a positive and negative way and therefore the writer and publisher must be sure that this is what they both want. These deals are common within the industry but when it's your copyrights involved, or your investment as a publisher, be sure you understand how the contract actually works in practical terms.

The best way to go through the complexity of such a contract is to revert to the earlier familiar terms and expand from there.

Territory

Thinking logically about the business transaction in this type of contract, the publisher will be committing larger advances to the writer and therefore it makes perfect sense that the publisher will want to protect its investment by making sure the writer is signed to them for the world. Therefore, most exclusive writer contracts will be for the world. The songwriter could have success in any territory around the world and in today's global economy international opportunities are very likely. A UK song could just as easily be covered by an American, Australian or Japanese artist and so the publisher will want to ensure that *all* income from *all* territories can be used to recoup against the advance provided. This is as it should be — this is, after all, a business deal. A worldwide deal helps to minimize the risk to the publisher, and provides an outlet for the writer's songs/music through the publisher or its affiliates worldwide.

Advances

Many facets will come into the negotiations, such as experience (is the writer a proven hit songwriter?), active covers (does the writer have artists recording and releasing his or her songs currently?) and whether the writer is a writer/artist (has a deal been offered by a reputable record label?). If a label is investing serious money into the marketing of an artist/writer album then the publisher will be quite excited by this and by the opportunity for a good return on investment. Of course, there is always a risk and in today's economic climate everyone is risk adverse. Another consideration is whether the writer is bringing his or her back catalogue into the deal as well (a writer with a back catalogue means there is going to be immediate income stream and hits that can potentially be remarketed for synchronization use and covers with new artists). Does the writer have some professional recording facilities? Is the writer a writer/producer (and therefore likely to have the opportunity of co-writing with other signed artists)? What type of music does the writer produce? If the writer is working in a niche area (drum and bass) as opposed to a broader mass market (pop/R&B) then the potential earning capacity is very different. Has the writer got more than one great song, is he or she a prolific writer, and does he or she have the ability to write many songs in a year and for them to be of a high standard?

As you can see, any combination of the above starts to build up tangible market value that the publisher can monetize. The 'added value' elements that the writer brings to the negotiations can have a significant impact when negotiating different points within the contract by the way in which these elements work with each other. I will try to explain.

Like all things in business, the stronger your negotiation position, the better the deal you can start with. I say this because if you do not have a lot to offer to begin with your lawyer can still provide for the likely event of it happening within the contract (at some point in the future). By way of example, your advance on signature of the contract may be lower than you'd like but the contract could be negotiated to provide additional advances based on future success. If success occurs the publisher will be happy to pay out some more money, and by leveraging the contract in this way it keeps the risk lower for the publisher but also fair to the writer. The second advance might have a minimum/maximum provision to provide additional money if the writer achieves above an agreed expected level of success. This would include a calculation of pipeline royalties (royalties earned but not yet received).

An exclusive writer contract will be broken into periods. The term may be for an initial three- or five-year period, but each year (contract period) will mark an amount of work the writer is expected to have completed, usually described as the 'minimum commitment'.

For example:

On fulfillment of the first contract period (usually based around a minimum commitment) the Writer will receive the next advance being £x amount plus a percentage of pipeline income up to a maximum of £y being £x amount plus 40% of pipeline income.

The whole point here is to remember this is a business deal; both parties have to feel that it's worth working together. The writer needs to be able to afford to work at writing full time and the publisher wants to make sure they are not too exposed to risk.

Two typical structures are used for advances. The first is on execution of the contract, followed by an annual based advance on the anniversary of the deal; but beware: this will only come to pass if the writer has fulfilled a minimum level of work (see below) in each year, the minimum commitment. Pipeline income can also be factored into the amount paid on the anniversary of the deal.

The second structure is a 'rolling advance'. There is a payment on execution of the deal and the next payment occurs when the writer has recouped the first. This structure hangs importantly on the publisher allowing a sensible provision for pipeline income, i.e. the writer may have had a No. 1 in the USA but not yet have received the income from it, which could take 12–18 months to arrive back in the UK. Therefore, a reasonable percentage (40–60%) of pipeline income should be allowed if the next advance, for instance, relies on recoupment of the first. There is usually also a cap irrespective of the amount of pipeline due. Or, perhaps the size of the second advance relies on the amount earned in the first year. Rolling advances typically favor the publisher as they reduce the financial risk to the publisher. If you are an independent publisher then rolling advances can be an easier way of building up your catalogue as the financial outlay is more manageable and provides greater control.

Retention period

In an exclusive writer contract the retention period typically follows the end of the term. If the writer signs a five-year contract, then at the end of the contract (completion of the term), the writer having first fulfilled all the necessary criteria for it to finish at five years, the copyrights are retained by the publisher, but the writer is free to leave and sign to a new publisher.

The contract will provide for a retention period on the copyrights to allow the publisher to continue working the songs, recouping their investment and making a financial return for themselves and further income for the writer. The retention period in today's market is likely to be anything from five to 15 years.

To clarify a key point – at the end of the term the writer leaves his or her existing songs with the publisher they contracted to, but can now start a new exclusive writer deal with a completely different publisher for any new songs to be written from this point in time. The writer is still accounted to by the original publisher in the normal way; a breach of accounting terms would jeopardize the contract and so the publisher will always look to account as per the contract to the writer. The writer has nothing to fear from the contract concluding and his or her songs remaining. If the publisher fails to account and they are notified (as per the terms of the contract) and fail to account, the publisher will be held in breach of contract and the songs would revert to the writer early, under the terms written in the contract.

As a publisher you would tend to draft the contract with the right to acquire any copyrights that revert to the writer during the term, but from a previous deal. If the writer does not wish this to happen then it can be excluded. However, if the writer needs to add weight to the negotiations then this can act as an attractive carrot to the publisher and is factored into the financial negotiations to help increase the value of the deal.

A typical clause would read:

If during the Term of the contract any copyrights that have been previously licensed or assigned to another Publisher revert to the Writer then they would form part of this contract and be considered as being assigned hereunder.

Negotiations

Consider for a moment that you are a publisher. You are about to commit to providing an annual advance of £20k, second year £30k, third year £50k (that's £100k investment). What minimum commitment do you want the writer to promise that they will provide to you? If you are the writer, what minimum commitment are you happy to give? Remember, this deal has to work for both parties. The publisher must feel that it will recoup its investment. The writer wants some freedom to develop and have hits. It's an important question for both parties. It is also a vital point to remember in business generally, that any successful negotiation is one that both parties feel good about. If you can understand the needs of each party, and what's most important to each party, then you can find areas to negotiate upon that will balance out risk and reward in a fair way. Let's explore this further together.

The way the publisher will recoup is all tied up in the minimum commitment; it is also the structure that can keep the writer tied to the publisher for perhaps longer than they feel comfortable. It's now about balancing risk.

Typical areas of negotiation

It has been decided that the writer must write 10 new songs a year (very reasonable).

The clause will go on to say 'written in whole or in part', i.e. has the writer written 100% of the song, so that all the monies earned can go into recouping the advance, or is it just 50% of the song because its been co-written with someone else? In this case only 50% of the income can be collected by the publisher, split between the writer and publisher, and the writer's share used to be applied to this deal. This is a crucial area of negotiation. If the publisher insists the contract says 'in whole' then if the writer does co-write then he or she may have to deliver 20 songs and not 10 in the same period, i.e. $20 \times 0.5 = 10$. How long will it take the writer to write 20 songs? If it takes more than a year the clock just keeps ticking, i.e. the year is not concluded until the minimum commitment has been fulfilled. The writer may decide that this is still absolutely fine. But what if the writer co-writes with three people? Then by the same formula there would have to be 30 songs. The writer and his or her manager must be realistic and look at how the writer generally works. One way around this is to ask the publisher to accept in whole or in part songs, suggest increasing the number of delivered songs (e.g. to 15) and add that any one song would not be less than 30% controlled. Balance what the writer can realistically achieve and what the publisher needs in return. There are some good points here to reread and reconsider carefully. Essentially, this mechanism is in place to do two things: to allow the publisher to recoup and to build a catalogue of songs from this writer. A further position could be negotiated: if the advance had been recouped (a hit in your own territory or a synchronization fee could provide this possibility, along with a percentage of pipeline) then the minimum commitment could be reduced in that period. The risk to the publisher will have been entirely removed, although it will not have succeeded in getting the volume of songs it may have wished.

The minimum commitment clause may also include that there must be a number of songs commercially released on a major label or major independent label, and it may state that such a release must sell more than x records to qualify. (The industry is currently experiencing vastly lower sales than in the past. A No. 1 UK hit might have generated 200,000 sales in the past, whereas now you'd be lucky to clock up 50,000 sales, and significantly fewer in some periods.) So what does a hit mean to a publisher in working out mechanical income? This is a reasonable issue the publisher has to consider, as the only way a publisher can recoup its investment is from royalties. Therefore, it will want a given minimum number of songs and some guarantee of releases and sales figures.

Of course I know your next question: how can a writer guarantee a number of releases? They are the writer! The 'release' addition to the minimum commitment will provide the publisher with a mechanism to drag out the initial period of time (a year) until such time as the minimum commitment has been met. The writer's lawyer will want to make sure that any one period cannot be extended longer than an additional two years. In law there has to be an absolute backstop to the agreement or it can be considered a restraint of trade. The publisher has to ensure, however, that the writer is committed and the deal will have every possibility of recouping. Remember, the publisher is taking a financial risk. If you don't need the money then get the best royalty splits you can and negotiate from a key area of strength. If you need the money then you need to consider both sides of the coin in more detail.

So, to be clear: a three-year contract could effectively be drawn out to last for nine years if in each period it takes the writer three years to complete the 'minimum commitment'. So how can this situation be improved for the writer?

Areas of negotiation may include an agreement that if the writer has recouped the advance the minimum commitment would be reduced significantly. The writer may be coming off a US No. 1, or have had a huge film synchronization fee (both are possible). In addition, the publisher may accept that what is deemed a 'release' may be broadened out to cover synchronized use of a song (often the advance is bigger than having a UK Top 10 hit). A television commercial, for instance, would provide a synchronization fee and ongoing performance income from the television performance of each advert featuring the music. This can be at least as lucrative as a UK Top 10 single. I see no reason why this could not be negotiated in such a way because of the increased income from synchronization licensing. Contractually, publishers have not accommodated this readily as they want to see a commercial release and radio to be a key component, but times have changed.

The contract should not be all about the writer — what about the obligations of the publisher? It is unlikely that you'll get the publisher to agree to a set amount of work. Having said that, Famous Music, the original publishing division to

Paramount pictures, used to do exactly that. Irwin Robinson and Alan Melina, who headed up Famous (I dealt with both) would agree to include a number of film synchronizations, and writers loved them for this. It was a great leverage for the publisher in their negotiations. These days the best you will get is likely to be the standard phrase 'the Publisher will use their best endeavors' to activate your copyrights. The relationship you have with your publisher is therefore really important. It is very much about building a long-lasting relationship where both the writer and the publisher work together to achieve mutual success. Again, consider life from the other side of the desk: if you're the publisher and you made some solid commitments and didn't keep them, your writers may be able to negotiate an earlier way out of their contract and you may be seriously out of pocket and lose your job for having negotiated such a bad contract. Again it's about communication, people and negotiations, and understanding what each party needs and is trying to achieve.

Royalties

This can also be a very complex area of negotiation and can hold many opportunities based on all the areas we have discussed so far.

- Performance income — Never less than a 50/50 deal. But a writer could achieve a split (in their favor) all the way up to 80/20.
- Mechanical income — The same applies to the above.
- Synchronization income — Generally most publishers will look for a far more generous split with the writer, as publishers may have to engage agents or others to help create activity in this area as well as having to engage their lawyer or business affairs to conclude the synchronization contract. More typically this is 50/50 to 70/30.
- Sheet music — A net receipts figure, as the publisher generally outsources sheet music sales, or it could be expressed as a percentage of retail.
- Covers — Performance and mechanical income on covers is usually lower by 10–20% than the headline rates agreed in the contract with the writer. It is usual for the publisher to expect to receive a better split on covers they or their sub-publishers acquire. For an independent publisher especially, operating with subpublishers is a way of incentivizing the subpublishers. The publisher provides the subpublisher with an extra percentage and the writer's share goes down by a similar percentage, but it is additional foreign revenue and again this is fair business.

Net receipts and at source

These are two terms that are absolutely crucial to understand, and which impact greatly on a writer's income. Both practices are commonplace in the industry, and apart from a small percentage reduction of income in one, both are good deals to have.

At source

All major publishing contracts will say that the writer is paid on an 'at source' basis. Very simply put, because a major publisher (by virtue of its definition) has offices in all key territories, it can collect directly in each territory and agrees in its contract with the writer that all income is paid at source from each territory detailed. The writer will receive their share without further deductions (after local society deductions, taxes and VAT where applicable).

For example:

Universal agrees a 70/30 'at source' split with its writer (on all income streams just to make this simple). The deal is signed in the UK for the world. The writer is guaranteed to receive 70% of all monies Universal receives.

If €100 is received by Universal Germany, €70 euro is the writer's share; this will track back to pass Universal in the UK and the writer will receive €70 (equivalent in sterling, no further deductions).

100 yen received by Universal in Japan: the writer is due 70 yen, this tracks back to Universal UK and the writer gets the equivalent of 70 yen.

There are no deductions: a 70/30 split means exactly that. Simple!

Net receipts

In the next section I continue to talk about subpublishers and subpublishing and the differences between an independent publisher and a major publisher. Well, 'net receipts' is a key difference.

Usually, independent publishers cannot offer an 'at source' deal. Not because they don't want to, or because they are trying to be dishonest in any way (not at all). Independents do not have offices all over the world. Therefore they have to engage (outsource) a third party agent to act for them in other territories. They can, however, offer an at source deal in their own territory and, depending on the split with the writer, this can also be achieved in some circumstances where they would pay an equivalent of an at source deal, the publishers swallowing the subpublisher's deductions. Clearly, this can only happen if the writer is on a 50/50 or 60/40 deal; after that it would be hard for the publisher to make a sensible profit if they were paying 20% to a subpublisher. On a 60/40 deal where the publisher would have 40% essentially this would then break down to 20% to the original publisher and 20% to the subpublisher. You can see that this is not something that most independent publishers will consider as standard.

In the same example as used above,

Tiptop Music (made-up name) has signed a UK writer and only has an office in London, England. They have engaged subpublishers or set up administration deals in the rest of the world. The money flow would work something like this. The writer contract will say that the writer is paid on a 70/30 'net receipts' basis. This is how it works.

The deal is signed in the UK for the world. The writer is guaranteed to receive 70% of net receipts from Tiptop Music, which has engaged Bigger Music in Germany to be its subpublisher. Bigger Music's deal allows them to be paid 20% off the top for their services (i.e. a 80/20 subpublishing deal with Tiptop Music in the UK).

€100 is received by Tiptop Music's subpublisher (Bigger Music) in Germany, who will take €20. Big Music will send back to the UK €80. Out of the €80, 70% will be accounted by Tiptop Music to its writer (because they have a 70/30 net receipts deal). 70% of €80 = €56.

Comparison

In an at source deal the writer received €70 and in a net receipts deal the writer received €56.

Legal position

There is absolutely nothing wrong with a net receipts deal and in all other respects it is conducted in exactly the same way as an at source deal. It is a method by which small independent publishers can still do worldwide deals with writers. The independent publisher is not benefiting financially over the writer, this is just how this mechanism works. If you're a small American publisher needing to grow your business, then you too are likely to pay your writers at source in your home country, but then outside of the USA, your contract will allow for a net receipts split so that you can engage a subpublisher in another country to act for you. You have to pay that subpublisher, and they will typically take 20% off the top as their fee (from all royalties they collect) and then remit the 80% back to you in the USA. You can then pay yourself and your writers on a net receipts basis.

Remember, a good publishing deal is primarily about the people you choose to work with: they need to have a sound business structure, be hard working and motivated, believe in you and be able to carry out all the back office work that is essential. There are many great independent publishers who do a fantastic job!

It is possible for an independent publisher to pay a writer on an at source basis, and there are two ways this might happen:

- in their home territory — this can be on an at source basis with their writers if they are doing their own administration work
- if the royalty splits with the writer mathematically allow them to do this.

For example:

A UK writer signed on a 60/40 for the world on an at source basis to Crispy Music. Independent has a deal with Peermusic for the rest of the world on an 80/20 basis.

Typical money flow:

- *Peer US — collects $1000 (takes 20% = $200) and sends back the remaining $800 to the UK.*

- *UK writer gets 60% of the $1000 = $600.*
- *Crispy Music gets $200.*

The writer in this example gets their full $600, i.e. an at source deal. Crispy Music ends up with $200 (i.e. 20% and not 40% because the subpublisher has been paid 20%).
A 70/30 deal with the writer would see:

- *the subpublisher taking $200.*
- *the writer receiving $700.*
- *the original UK publisher taking only $100.*

Therefore it starts to become an unviable financial proposition for Crispy Music. Crispy would not be able to do an 80/20 deal with a writer on an at source basis with this structure as they would end up with zero profit. This is why net receipts deals exist.

Co-publishing Agreement

Co-publishing agreements are far more common in the USA than in the UK. The songwriter and the music publisher co-own the copyrights of musical compositions governed by the agreement. A split of the royalties is agreed where the songwriter assigns a percentage to the publisher. The split is normally (but not always) 50/50, with the writer assigning the publisher's share to the publisher, and retaining the writer's entire share. In the UK a writer will rarely get only 50% of the copyright, i.e. just 100% of the writer's share. The minimum is generally upwards of 60% of the copyright, i.e. 100% of the writer's share and 20% of the publisher's share. It makes for good sales language if you can say to a writer, 'you'll get 100% of the writer's share', doesn't it? You can see why this is a preferred way of describing matters in the USA.

In a 75/25 co-publishing deal the writer keeps 100% of the writer's share, and is given 50% of the publisher's share, which is 75% of the entire copyright. The remaining 25% is the publisher's share (and profit). When royalties are paid the writer will receive 75% of the income, while the publisher will receive 25% of the income.

In the UK we do not refer to the writer's share being 100% and the publisher's share being 100%. Who administers the deal is important. With only one copyright owner this is obvious, but when copyright ownership is shared the exact roles and responsibilities need to be defined. The same is true when there are any restrictions placed upon administration, or when moral rights are asserted.

My experience with co-publishing agreements has been rather different and you will find that this particular contract may be explained in several different ways by different lawyers. Many years ago a well-known writer whom I had known and worked with in the past approached me. I had set up various collaborations for my husband and I chose this writer for him to co-write with. The songs that were written all got cut and everyone was very happy. He had approached me feeling dejected as his own publisher had not produced as much activity for him as I had (by default).

He wanted me to sign him but wanted a sizeable advance and at source accounting. I couldn't supply either. Troubled by this, I pondered for a few moments and decided what was needed was a co-publishing arrangement. I wanted to partner with another publisher just on this writer; they would pay the advance and take care of all the accounting but we would agree to split the publisher's share. To my delight EMI agreed the deal. I was the 'active', creative publisher within the deal, EMI supplied all the administration, advance and collection structure, and my role was purely A&R, but I received one-third of the publisher's share with no accounting deductions and no risk. So now you have two quite different descriptions of a co-publishing agreement — a writer signing to a publisher under a co-publishing agreement and two publishers representing one writer as a co-publishing agreement — and there are several variations on both.

Administration agreement

This is the simplest of deals and is a business-to-business (B2B) deal. It is generally used between one publisher and another publisher. Typically, a small publishing company wishes the collection of its copyrights from around the world in as cost-effective a manner as possible. Often this concerns an experienced writer who has their own catalogue and perhaps their manager or the writer taking care of all their own procuring of activity. Usually no advances are paid in this regard and the fee charged by the administrator for carrying out this job is by way of a percentage (typically 10–15%) of the income received. Usually the administrator is granted a licence for a period of time as opposed to an assignment of rights. An administration deal can be attractive as it allows the copyright owner to engage one administrator per country or one for the world or any combination thereof. Total flexibility is available if the owner so wishes. The owner will build up a B2B relationship with the administrator and not a creative relationship.

Subpublishing contract

This is usually a publisher-to-publisher deal. The publisher could, however, be the original catalogue owner, i.e. the writer who has set up their own publishing identity (my husband's is G2 Music). The idea is that the local publisher looks for someone to represent its catalogue in another territory. The subpublisher should have a number of qualities that you are in need of such as experience and proven success in other markets, knowledge of their territory, advances, creativity, good staff and administration capabilities. The subpublisher could be a major publisher or a significant independent publisher in that territory. You should seek a publisher that can bring value to the relationship.

A typical contract may be for three to five years and generally an 80/20 deal is considered a sensible split. What does this mean? Imagine that you are based in the UK or USA and you need representation in Australia. Owing to songs being aired on television in one of the many soaps you want a subpublisher to collect those incomes for you and to maximize your success by getting other songs used, if someone in that country was working your catalogue. This is when you need a subpublisher. They will collect *all* income streams, 100% of mechanicals (from the sale of records), 100% of local synchronizations, and all available performance income (usually only 50%), the writer being paid 50% directly via membership of their original collecting society who will, in turn, have reciprocal agreements around the world.

In the UK, PRS for Music members will receive 50% of all performance income from all countries around the world under reciprocal arrangements with other societies. So what does 80/20 mean? Simply, it means out of all the monies the subpublisher could collect in a given accounting period (and let's just assume that this amount is $5000) the subpublisher will keep 20% ($1000) for doing that job and send back the balance to the original publisher (being $4000). Then, based on the contracts the original publisher has with its writer it will account the relevant percentage onwards to the writer in the next accounting period.

The original publisher could elect to have a different subpublisher in each territory if it so wished. I will discuss the pros and cons of this shortly. The other way of dealing with the representation an original publisher may need abroad is to go to a major publisher and do a deal where they subpublish the original publisher all over the world except for your original territory. Of course, every combination in between is also possible.

There is no right or wrong way of dealing with this. It's about the strategy you want to follow and the vision you have for your business, what you want to achieve and how you want to achieve it. These are very big and bold statements to make. Let me help you here by discussing the pros and cons of any decision making.

SWOT analysis — going to a major publisher for representation worldwide as a subpublisher

Strengths

- You would potentially receive a larger advance.
- There would be only one contact to update with your titles as they will disseminate around all their offices.
- You will have access to top writers and artists.
- The publisher will have dedicated synchronization departments.

Weaknesses

- You would not have an existing relationship or have even met the different MD or A&R staff in each territory. How would you know whether they liked your music in Spain, France or Germany?
- The bottom line is key: will your small catalogue receive attention?
- Should the person who signs your deal leave, you will effectively have no contacts within the organization.
- The advance paid is cross-recoupable from all territories, i.e. it doesn't make any particular territory work hard for you.
- They handle a large volume of copyrights with few staff.

- Their legal affairs departments are slow to react to synchronization uses. A $2000 deal will not have the same sway as a $30,000 deal for their primary copyrights.

Opportunities
- Working with just one company will free you up to focus on what you want to be doing.
- If you go to them with significant success they are well placed with larger departments to secure future activity.

Threats
- It might be good on day one, but in 18 months' time will you still want to be in the deal if no one is talking to you from around the world and you feel that your songs and catalogue have been forgotten?
- Mergers and acquisitions has left its mark. Supersized major companies have fewer creative staff than ever before, something has to give, and it's usually creativity and the smaller (more time-consuming) catalogue owner.
- Slow reactions to synchronization use by legal affairs could lose you money. (I've seen it happen many times.)

SWOT analysis – going to an independent publisher for representation in different territories

Strengths
- The advance paid in one country is *not* cross-recoupable from any other territories, i.e. every territory has to work hard for you.
- You are often dealing with the owner (MD) and have security in knowing who you will be working with for some time.
- Smaller companies are agile and more entrepreneurial than major companies. They can respond to market changes more rapidly.
- Every penny counts and so synchronization contracts are dealt with more quickly.

Weaknesses
- There is a patchwork quilt of companies that you will have to stay in touch with and update accordingly. This takes up a lot of time.
- Your own administration workload is higher, including having to input data from a number of accounting sources back to your writers.

Opportunities
- Attending international conferences will assist you in securing the best local publishers in each territory, getting companies in place who want to work with you and help you grow your success. Local knowledge is very important.

- Advances from each territory are not cross-recoupable. Each territory must work hard to recoup any advance it has given you.

Threats

- Smaller companies may have some financial risk attached to them, especially in these turbulent economic times. So check with your lawyer and other sources as to who is reliable and strong.
- More time spent on administration may detract from other things you want to do.

My original structure, which I shared in an earlier chapter, was a combination of major and independent publishers (best of both worlds) but I did elect to go territory by territory and for me this was a significant success! Personally I think the only reason to do a major worldwide deal now is for the size of a single advance and to cut down on administration. If you are a dedicated independent publisher then there is significant merit going territory by territory; if you are a writer owner wishing to continue writing as opposed to working the catalogue then you fall between a rock and a hard place. What you need is less administration, but you do need a creative relationship. Maybe then licence or assign your catalogue in chunks, dealing with the USA as one territory, Japan and the Far East, and then Europe. You would end up with three or four subpublishers. If it's creativity that you want then you'll have to split up your deal somewhat, otherwise you will never successfully build your network or get enough creative support in key territories.

My own network is very well developed now and I have had the privilege of working with many wonderful people. This is the crux of the matter. When you do a deal you're doing a deal with people and they must really buy into your music, your catalogue and what you want to achieve. Let that guide you as to whom you should ultimately work with. I've had a long-lasting relationship with many of Peermusic's MDs around the world (being a major independent they often bridge both worlds, independent and creatively minded but with a strong administrative structure and people I trust). But I know and love the work of many other MDs from around the world who work for the majors — EMI, Universal, Sony. I have found some territories very challenging. France and Japan are two countries that need careful consideration. I have encountered cultural issues and different working practices along the way.

Collection agreement

This is similar to an administration deal in that the catalogue owner licences the copyrights, while the publisher does not have any creative or exploitation role. The publisher only collects and accounts earned royalties governed by the contract.

Synchronization agreement

This is dealt with in Chapter 7.

SUMMARY

Many different contracts are used in the world of music publishing. There are, however, core terms and language that will be familiar and recur from one contract to the next. Becoming familiar with these will allow you to build your understanding and confidence in being able to navigate through this world. You need to be able to understand the key commercial points of any deal, its structure and its relevant impact on you. Carrying out a SWOT analysis may be a good tool to explore the pros and cons of a given deal for you at any given time. Again, I stress: always use a good lawyer for any contract.

Synchronization

Let's start with a basic definition of music synchronization — it's the ability to synchronize music in a timed relation to a visual component. As such, you can now begin to appreciate that it's not just movies we're discussing but everything in the above description. This includes computer games, television shows, film, advertising commercials (television, cinema, online, video). It will cover all new devices that allow this to take place and in any new medium.

Up until now synchronization income has shown hardy resistance to downward trends. Indeed, new avenues of synchronization have brought new income streams and helped music publishers (and record labels who also have to give permission for the use of the master recording in synchronization use) to secure relatively substantial incomes. This has helped to compensate for the decline in mechanical income (and labels' record sales). So a success story all around! The financial pressures on the film industry in 2010 saw MGM Films, which owns the rights to the *James Bond* intellectual property (IP), get into major difficulties in fact they have now gone into liquidation. A pre-packed bankruptcy plan saw MGM merge with Spyglass Entertainment, whose co-CEOs will lead the combined company. The Film Industry is having a torrid time of late with film finance generally in freefall. More tax incentive packages are needed especially if the creative industries are to be the instigators of growth in both the UK and American economies. Synchronization income is likely, in my opinion, to come under some pressure and money paid for synchronization use will diminish, with smaller budgets and less money for music clearance and commissioning. There is evidence that film and television companies together with advertising agencies are already paying much lower fees. Some primary research I conducted across independent and major publishers was conclusive in this regard. Some surprising full-year results released for Warner Music Group show only a 2% increase in synchronization income in the year but the last quarter being significantly lower, by some 25%.[1]

Synchronization income has to date largely bridged the gap created by the decline in mechanical income. It's important to both the writer and publisher to understand all that you can about music synchronization: what it means, how income is generated, what markets can be developed, who the players are and how to operate within it. You must not be blind to market forces but negotiate and secure deals that put your writers at the front of the queue for that important job. A political, economic, social and technological (PEST) analysis will help you to gather information; review

---

[1]http://www.wmg.com/newsdetails/id/8a0af81224c70dce012523c2bd396ae9

The Art of Music Publishing. DOI: 10.1016/B978-0-240-52235-7.10007-7

what is happening in the market, looking at press, statistics, competitor news and gossip, and you may learn something that may give you an edge.

Identifying who your customers are likely to be, and how their needs and wants should be addressed, is very important. This has often been the stumbling block for the music industry, preferring the attitude if you want it, come and get it, but on our terms only. We cannot risk this attitude any more. These are challenging times and good negotiation skills are important. I don't mean just lower your price, I mean getting more tracks into a film or game and working on cross-platform marketing that may assist everyone. I also mean proactive support (some old-fashioned joined-up thinking and music publishing legwork). Many staff employed by music publishers fail to really listen to what the customer wants.

For example, I've been in the offices of NBC in California where senior executives have been bombarded by CDs from major publishing houses after letting them know that they are looking for a specific type of track. Instead of the publisher doing their job properly they merely sent over a whole stack of CDs for the executive to waste time going through. That's not good! If you want to be invited back and to have the ear of the executive then you need to do your homework.

This picture is repeated in every office around the world. There are some very lazy A&R people out there.

Some very forward-thinking independents stepped in and have made a big splash in music synchronization. Not only did they respond quickly and accurately, but they also empowered the film or television executive by giving them access to the publishing catalogue 24/7. Using intelligent search criteria, this was a brilliant idea and an immediate success. One of the first independent executives pushing this idea was Paul Kennedy. At this time he was working for Big Life and, spotting what he was doing, I cornered him for a chat, to find out who I could engage to do the same for a company I was consulting for. In 2003/04 we were able to search by tempo, work, lyric lines, style, mood, genre, instrumentation and many other more subtle parameters as well. Providing this service was fantastic. Executives were using the service at all times of the day and night and back-end metrics proved invaluable.

This search engine service still went hand in hand with selective presentation of tracks. It was the independent market that drove this and now all publishers have such facilities; they are commonplace and taken for granted somewhat, but it was the independent publisher who spotted this opportunity first.

I've had the pleasure of working alongside a number of companies who do this job very well indeed: Pig Factory, Music Supervisor.com, Leap Music, to name but a few. What I and these other companies were doing was offering 'one-stop shop' solutions. We found out what the creative executive wanted to use and then took care of all clearances for them or directed them towards titles that were already pre-cleared. What this did was speed up the process, which was essential as often programs had to be cleared within 48 hours. There was a clear need and the demand was primarily filled by entrepreneurial music publishers seeing a gap in the market: a gap the majors just couldn't respond to.

I will be launching a new service in due course around music synchronization and film funding — you'll have to watch this space.

In stark contrast, the recording industry's inability to negotiate synchronization use and its hostile attitude forced many telecommunications companies and computer games companies to set up their own publishing companies, labels and distribution platforms for content. In much the same way as consumers were pushed aside, now the media giants were also being boxed into a corner. In general, when someone is boxed in, it focuses the mind and they come up with solutions quickly and come out fighting.

This lack of foresight by the bigger players allowed the independent sector to walk in and seize this opportunity again. Like many changing industries the early adopters were the independent market. The independent market was quick to see and appreciate the opportunities that new outlets for their music could provide. But for these massive media companies content and 'prime content' (named artists, big hits and quantity) were uppermost in their minds and so the push towards working with the major labels and publishers still had to be very much on the agenda, and has been relentless. Out of this constant frustration, their needs to acquire rights began. Artists and their managers leapt towards such opportunities. Brand name artists (heritage acts) that were no longer wanted by the mainstream music industry found grateful new partners. New competitors had gained a foothold in the industry. You will find that many media companies now have their own labels, and publishing divisions commission work to be created for them by artists that they own. Keep this point at the back of your mind.

Hmm … media companies and mobile telecommunications companies have worldwide distribution, with a customer base far in excess of that of the music industry, plenty of money and a growing market. What they need is content. You figure it out!

With advances in technology new opportunities always open up in the market allowing new business opportunities and entrepreneurs to flourish, and this will continue. Look for such opportunities. It's seeing these gaps and responding quickly in a manner that can be profitable that makes the music industry and music publishing in particular so exciting. Be part of the changing landscape and not a casualty of it. (Have I said that before? Probably, but it's important.)

EXPLOITS IN FILM AND TELEVISION

What will a company pay for using your music?

The value of the rights that are required for clearance is broken down into bite-sized questions, the answers to which help to price the music.

This includes consideration of:

- **Setting** — Is it a major blockbuster, an independent film or local television? They could all use the same piece of music but what they pay is very different.
- **Duration** — How much music do they want to use (seconds, minutes)?

- **Specific use (buyout)** − What type of use is the music being put to? It may be the opening titles to a television series being screened weekly in the UK but being sold or transmitted worldwide; a television commercial in the UK that will run for six months with options for Europe and the USA; or music placed in a movie to be cleared 'in perpetuity', i.e. the movie is a finished stand-alone product and music cleared for use is embedded once and forever then screened in theaters, sold on DVD and screened on television, and this could go on for years. The music is cleared one time, for all use and forever.
- **Position of the music** − Is the music key to the experience? In the opening credits or end credits? Or a significant moment in the movie when the music is the feature and not buried under dialogue?
- **Added value** − Does the music add value in terms of being a well-known song?
- **Commissioned work** − An approach is made to a writer to compose specifically for a given use. Any negotiation has to be in two parts. Money is paid for both the music (composition, generally handled by the music publisher on behalf of the composers) and the use of the recording (master recording negotiated generally with the record label). Film companies or music supervisors will often say the fee is £10,000 all rights. They would be referring to the amount paid in total for both the recording and music clearance to all parties for the rights they need.
- **Music libraries and catalogues** − Music publishers are developing library or production music catalogues from which television companies can use their music in exchange for a subscription fee and/or blanket fee for the use. The publishing companies rely on the performance income for future payments. This has proved a good business model to date. It allows musicians who also compose to write and record their own material, thus controlling all rights (owning the complete pie). Developing your own music library catalogue is a great way of starting your own music publishing business. If you Google music catalogues, music libraries and blanket fees you will see a huge list of companies offering such facilities.
- **Work for hire** − Legitimate work for hire involves a composer seeking to be hired for a project. The composer is paid and delivers the work. The writer retains the writer royalties and the employer becomes the publisher. This is commonplace in both the UK and USA. Many UK television companies have their own publishing divisions and not agreeing to sign to them may lose you the job. Remember, television companies have to pay performance income fees and so if they are the publisher of the work they are effectively clawing that money (the publisher's share) back.

  Illegitimate work for hire involves the same situation as legitimate work for hire. However, sometimes the employer asks the writer to remove their name and give up the writer royalties. Sometimes this could even be a ghostwriter to a household name writer who has engaged someone on such a work for hire basis. Both practices are commonplace in the USA, and care and attention need to be paid to any such contract of engagement.

PRS for Music offers a scheme to assist publishers who wish to offer their music under blanket licence, if they want to opt in to this. They set the following criteria:[2]

- What is the name of the scheme?
  The Independent Production Company (IPC) Blanket Licence.
- Who is it aimed at?
  The IPC scheme provides a simple and cost-effective way for producers of television programs to access your music, without requiring individual approval from the copyright owner.
- What are the key benefits?
  More money for you — ease of use and competitive rates will lead to more production companies using your music.
  More efficient — no need for you to negotiate directly with production companies, giving you more time to market your works.
  Less cost — removes administration and debt-chasing costs.
- What does the IPC scheme cover?
  Production music (music that is specifically written for inclusion in audio and television productions).
  Commercial music (all other music).
- What does the scheme exclude?
  The IPC scheme does not cover the following usages:
  International use (it is for UK programming made by UK IPCs only).
  Adaptation, alternation, sampling or arrangement of a member's commercial music without prior approval.
  Reproductions of music as parodies.
  Use of music in deliberately derogatory, obscene or demeaning ways to its composer, author or the performing artist.
  Advertising or sponsorship.
  Use of commercial music in title/credit sequences.

BMI and ASCAP both have blanket licence rates and download contracts available on their sites. An interesting article appeared in January 2010, which BMI is fighting.

The owners of approximately 1200 local television stations, citing a 'massive reduction' in audience size, have sued performing rights society BMI in rate court, asking the Court to set 'reasonable fees and terms' for the use of BMI music in television programming including terrestrial and digital platforms, grant broadcasters access to cue sheets, and add a provision to the blanket licence that essentially creates a credit for each performance of a musical work that is directly licensed with a composer or publisher. The terms of the new blanket licence credit, if approved, are likely to be decided by

---

[2]http://www.prsformusic.com/creators/membership/MCPSroyalties/mcpsroyaltysources/broadcasting/IPC/Pages/IPC.aspx

a judge in a case brought by BMI against DMX, a commercial background music company.[3]

Universal Music has the largest music library:

Production Music Online UK

Universal Publishing Production Music is a diverse and comprehensive collection of over 4500 CDs with over 70,000 titles. Our renowned labels include Atmosphere, Bruton, Chappell, Match and many more and we cover every genre of music from live epic orchestra to solo instruments and underscores to pop/rock. They are a production music library offering a full music supervision service to our clients including free music searches, bespoke music edits, re-versioning service plus commissioned music.[4]

THE ROLE OF THE MUSIC SUPERVISOR

Music supervisors are found mainly working in the film industry, but with growing areas of synchronization opportunity their work now spans all areas of media. They work and interface with all parties involved in the provision of music, finding new music talent across any genre.

A music supervisor may also play a key role in establishing and managing the overall music budget and music production for a project.

Music supervisors negotiate deal points and contracts, prepare budgets, and attend scheduling meetings and spotting sessions. They oversee the compositional process, ensuring that the required music is being written, listened to and reported upon. When organizing source music, music supervisors prepare source music schedules and keep everyone informed and updated, e.g. about deviations from allocated budgets. Music supervisors check licences and forward them to the film production company, highlighting any possible issues, and act as the liaison between the record companies, the publishers and the production company. They may also produce the music cue sheet for final delivery, ensuring that the duration of the music used conforms to the terms of the negotiated contract.

Rates of pay for music supervisors vary depending on the size of the task and the weight of music and source cues needed to be cleared. When budgets are squeezed (and they often are) they have to get very creative with the budget they are given (or that which is left over after the movie shoot).

They need a great international network to call upon for favors. Often the music budget is the first to be trimmed if the movie has not completed on budget. The music supervisor needs to have a great knowledge of music — current and historical.

[3]http://www.filmmusicmag.com/?p=4856 By Mark Northam
[4]http://www.4rfv.co.uk/directory/193x7231/Bruton-Music-London.htm

Very few film directors, especially early in their careers, really appreciate the importance of music.

SOME KEY MOVIE TERMS

- **Spotting a movie** — The spotting session is when a director and composer get together to watch the rough edit film and decide where the music is going to go and what it's going to do. This occurs before the composer starts writing the music.

The composer will require a 'locked edit' of the movie to work to as until then the music and audio cannot be in timed relation to the visuals and you would not be able to orchestrate against anything other than that. However, the composer may well be working on musical themes and ideas well before this stage.

People present should be the director, the composer, the music editor and perhaps the producer and music supervisor. If the movie is heavy in source cues then the music supervisor's input and involvement will be both helpful and critical.

- **Sound design** — This is not an area in which the music supervisor will usually be involved.

PIERCE BROSNAN AS JAMES BOND

I have acted as music supervisor on a few projects. *Within the Rock*, starring Xander Berkley (*Apollo 13*), was one such project. I enjoyed the experience immensely, especially going to the American Film Market (AFM) to watch the first screening to the buyers. It was incredibly moving (being so proud of my husband and the work he did in doing such a great job in just four weeks from start to finish. I shed a tear or two; very emotional. We also published and owned all the music). Working on a movie takes a huge amount of time, and since then I have preferred to work on placing source cues for synchronization work (on a track-by-track basis). I have placed hundreds over the past 20 years and it fits in well with all the other work that I do.

My next project has just started. In 2010 I have been engaged as executive producer to conclude raising the additional funding required for a new Pierce Brosnan movie. As a consultant to the movie this is also very exciting and allows me to tap into another area of my network, built up over the years.

However, the story I wish to tell goes back to a time around 1995/96, when the James Bond movie *Golden Eye* came out. Pierce Brosnan was touring the globe doing some heavy film promotions and one country he visited was Japan. At this time I headed a music publishing company as creative director of Prime Direction, owned by Avex Inc., in which I was also a shareholder. That night the Japanese management team (some 30 or so people) and I were out celebrating some major work achievement at Velfarre, a nightclub owned by Avex. Low and behold, in walked Mr Brosnan, Bond himself. He was immaculately dressed and you couldn't help being impressed as he wore his classic Bond attire and was escorted by two burly minders. I nudged the CEO and pointed out who was in the company's club – James Bond (Bondo San) – and suggested that they provided a bottle of Dom Perignon to their distinguished VIP, which he did. It was gratefully accepted. One of the Japanese managers asked Bond if he would kindly oblige and have his photo taken. He courteously agreed, but as he slowly rose and before he'd even straightened his legs a queue had formed around the room of some 30 Japanese staff, all waiting to have an individual photo. I was mortified, can you imagine … I was red with embarrassment, but it was also so funny, I was laughing my socks off inside, but … I also wanted a photo! As the last person finished with Bond, I jumped up and shook his hand, introducing myself, and apologized profusely. We had a bit of a chuckle; frankly I think he was relieved to find someone he could talk to. I explained how innocently I had suggested they provide him with a bottle of their best champagne. He laughed again and then whispered to me (as all the Japanese staff still gathered round and continued taking shots), "shall we give them something to snap?" At that moment I saw an Irish twinkle in his eye, my heart missed a beat or too and having a mischievous moment I enthusiastically agreed. A rather posed kiss took place (I have the evidence, ladies) – what a fun memory. Now, some 15 years on I might be responsible for one of his up-and-coming movies! (Just no one remind him, OK? Our secret.)

So, back to the chapter ….

MOVIE: *WITHIN THE ROCK*

As music supervisor to this movie (see above) and working with a miniscule budget I had to pull as many favors as possible to enable the score, source cues and soundtrack album to be done. It is possible in such a situation to ensure that all rights, save for clearance and use by the movie, are retained by the composers and thus all other income would fall outside the synchronization licence negotiated with the film company. The film company did not end up owning any rights to the music save for a licence to use the music in the film in perpetuity. And so, the movie was made, supported by Prime Direction, the composer's publishers, 10 source cues, an orchestrated underscore blended with programmed work setting the scene for a fantastic blend of rock and industrial metal flavors and a Sci-Fi premier on Sky Movies. This was so much fun. A big thank you to Graham Bonnett (mix engineer extraordinaire who worked 24/7 with us for a week during mixing).

- **Contracts** – The most important contractual element in synchronization use is to ensure that it is 'non-exclusive', so that the music you place in a particular 'use' can be used in another movie or game, as you wish. Some typical wording might look like this:

  The Composer hereby licences to company and its assigns, on a non-exclusive basis, the following Rights including, without limitation, the complete, unencumbered, exclusive and perpetual right throughout the world to exhibit, record, reproduce, broadcast, televise, transmit, publish, copy, print, reprint, vend, sell, distribute, perform and use for any purpose, in connection with the motion picture as defined herein, whether or not now known, invented, used or contemplated, and whether separately or in synchronism or timed relation with the picture or trailers, clips or portions thereof.

In computer games you may wish to make sure that there is a repeat fee or that you limit the rights granted so that it doesn't include a series of games sequels and prequels unless additional fees are paid either at the time of entering into the contract or as they are requested.

The Top 20 Games Publishers in 2009 were:[5]

1. Nintendo
2. Electronic Arts
3. Activision Blizzard
4. Ubisoft
5. Take-Two Interactive
6. Sony Computer Entertainment
7. Bethesda Softworks
8. THQ
9. Square Enix
10. Microsoft
11. Konami
12. Sega
13. Capcom
14. MTV Games
15. Namco Bandai Games
16. Warner Bros Interactive Entertainment
17. Disney Interactive Studios
18. Atari
19. Atlus
20. Lucasarts

YouTube has some great games clips; go and check them out.[6]

- **Credits** — Always insist on both games and films that the publisher and composers receive a credit (protect everyone's moral rights). Where possible ask to check the credits and sign off on the wording so that they are correct.
- **Title song** — If you supply the title song to a movie try to negotiate a front- or back-end credit where you get a card (screenshot) of your own (no one else featured on the screen at the time) or perhaps a credit featured with the artist performing the song.
- **Composer of the score** — Many of the featured composers below will have a front-end credit to the movie as their name carries weight and added value. It may not happen if the movie doesn't feature credit titles at the front (some don't), but as the movie ends a featured credit should follow.

I'm one of those people who sit to the end of movies and read the credits to see who I know, and who's been doing what. If you are involved in the making of a movie ensure there is a credit for your work and/or a 'thanks' at the end.

---

[5]https://store.cmpgame.com/product/5612/The-Game-Developer-Top-20-Publishers-Report-2009&skin=gdmag

[6]http://www.youtube.com/watch?v=DMTDjKusVqY&feature=player_embedded
http://www.youtube.com/watch?v=u17DloxYmzY&feature=player_embedded
http://www.youtube.com/watch?v=9rbeAGdYk_0&feature=player_embedded

INDEPENDENT FILM SALES NETWORKS

- AFM (The American Film Market) — Los Angeles — February — Buying and selling movies. A very exciting focused sales convention. www.afma.com
- Cannes Film Festival — May — Buying and selling movies. The premier location for screenings (same place as MIDEM). Very exciting, harder to do business, 10 times as many people attending. www.festival-cannes.co
- Sundance Film Festival — January — http://festival.sundance.org
- London Film Festival — October — www.mofilm.com

GREAT MOVIE COMPOSERS

- Bernard Hermann — *Psycho, Vertigo, The Birds, Citizen Kane*
- John Williams — TV shows: *Lost In Space, Time Tunnel* and *Land Of The Giants*; films: *Jurassic Park, The Patriot, Jaws, The Towering Inferno, Star Wars, Close Encounters.*
- Gerry Goldsmith — *Total Recall, The Mummy, The Sum Of All Fear, Star Trek Voyager, Planet Of The Apes, The Man From Uncle, The Omen, Logan's Run, Air Force One, Alien, Star Trek The Motion Picture, First Blood, Gremlins, Sleeping With the Enemy, Basic Instinct.*
- Hans Zimmer — *Gladiator, Hannibal, Pearl Harbor, Mission Impossible (sequel).*
- James Horner — *Braveheart, Apollo 13, A Beautiful Mind, Titanic, Avatar.*
- Danny Elfman — *Men In Black, Batman, Planet Of The Apes, Spiderman, Good Will Hunting.*
- Don Black OBE — Oscar-winning lyricist: *James Bond: Thunderball, Diamonds Are Forever, The Man With The Golden Gun, The World Is Not Enough.* In collaboration with John Barry he wrote the title song to the 1966 film *Born Free,* which won an Oscar for best song. He also collaborated with Barry on *Dances with Wolves* and *Out of Africa.*
- David Arnold — a Grammy Award-winning English film composer, best known for scoring five *James Bond* films, the blockbuster *Independence Day* and cult television show *Little Britain.* He has worked on various soundtrack projects, including *The Young Americans, Stargate, Last Of The Dogmen, Godzilla, The Musketeer, The Stepford Wives* and *Four Brothers.*

In 1997 Arnold produced *Shaken and Stirred: The David Arnold James Bond Project,* an album featuring new versions of the themes from various *James Bond* films. Away from the film world, David Arnold maintains a career as a successful record producer and songwriter, having worked with a wide range of contemporary artists including k.d. lang, Pulp, Chrissie Hynde, Iggy Pop, Garbage, Massive Attack, David McAlmont, Martina Topley-Bird, Natasha Bedingfield, Aimee Mann, George Michael and Damien Rice.

CUE SHEETS

The synchronization contract is signed, the film is made. As publishers you now need to attend to collecting your income stream. The cue sheet is the cornerstone of all royalty payments for a film or television show. Considering the amount of music used in most films, the cue sheet is usually finalized within 30 days after completion of the movie. However, sometimes this job falls to the music supervisor to ensure everything is listed and accurate.

Failing to file cue sheets will result in no money reaching the publisher or the writer. Filing cue sheets is an essential part of synchronization work!

USEFUL FILM BODIES

- The American Film Council — www.americanfilmcouncil.com
- The British Film Commission — www.bfc.co.uk
- UK Film Council International (UKFCI) (which may now be disbanded by the UK government)
- www.bafta.org

THE TOP UK AND US ADVERTISING AGENCIES

How do you get your music to ad agencies?[7]

In short — you need to be creative. You can try the direct approach (very hard), or perhaps try going to the brand direct. Find out which companies they represent and for which products. Your lawyer, accountant, even bank manager may be a good source of assistance. Music industry organizations such as the Music Publishers Association (MPA) and the Music Managers Forum (MMF) may also be able to assist in making introductions. The BPI is actively involved in helping labels build synchronization contacts with advertising agencies, games companies and film company executives. The advertising arena changes very quickly, with account executives moving from one agency to another fairly rapidly. Remember, music is just one part and probably a small part in their minds in the overall advert and pitch they are making to brands and companies, so again, you need to understand life from their perspective. You will always think your music is best for them, but try to describe the music in a manner that may fit with the brief. They are often looking for an emotive link, a psychological interplay between music, brand and demographic and, of course, increased sales of the product.

All the companies below have UK offices but many also have offices worldwide. However, this link provides a brilliant list to all the American advertising agencies.

---

[7]http://www.artistshousemusic.org/videos/getting+your+music+to+ad+agencies

American agencies:

- http://www.adbrands.net/us/index_agencies.html

UK and American agencies:

- McCann-Erickson — www.mccann.com
- MC Saatchi — www.mcsaatchi.com
- J Walter Thompson — www.jwt.co.uk
- TBWA — www.tbwa.com
- Abbott Mead Vickers — www.amvbbdo.com
- BBDO — www.amvbbdo.com
- Saatchi & Saatchi — www.saatchi.com

SUMMARY

I hope by now you are really beginning to see an exciting picture forming in front of your eyes. Music has so many uses (aren't we lucky?). A song can be an artist release, a movie or television synchronization, or it can have the vocals stripped off and just be used as an instrumental track (library use). The lyrics could be used in a portfolio or by a talking, singing clockwork toy. Music can come to life in a computer game or a children's animation, or be featured in a major blockbuster or a small independent film, as a source cue or underscore. Music can enhance a brand and product and it can open new markets. Music can be a physical product or a digital file, a ringtone, a real tone, a karaoke favorite, sung in the back of a Japanese taxi. Music you wake up to in the morning on the radio or the last thing you hear on your iPod at night, or the event of the year at a festival. How exciting is all this? All of this creativity requires songwriters, music composers and music publishers to administer. This is why we must protect them, protect their art and make sure the future of music and those that create it is long lasting.

Getting a publishing deal

8

Have you jumped straight to this chapter? OK, that's fine, but you've missed the chapters on where the gold is buried and how to make your millions, so go back afterwards and read from the top. Without a working knowledge base of all key aspects of music publishing you will never be totally in control or feel empowered to develop your career as a writer or publisher.

On this very note, Camilla Kerslake (the first artist signing to Gary Barlow's label) has said that her time spent on a Music Business course was the best thing she has ever done, and lawyers she met to discuss the terms of the artist contract were impressed at her knowledge base. As the old adage goes, 'knowledge is power'.

I do understand that those who are creators just want to be getting on with creating, but music publishing itself is a very creative environment, it is a creative way of life. I don't personally write music, but I know I bring a lot of creativity to people's lives. That's my skill, I just can't copyright it ... well, there is this book!

I'd like to bring in a quote here from Eric Beall, from an interview conducted by Michael King on www.artistshousemusic:

> *The biggest misconception I see about what songwriters think about music publishing is that they see it as an 'Us and Them' arrangement. They think of songwriting and publishing as two separate things. And the fact is, as soon as you write a song, you just became a publisher. You are the publisher of your song, unless you decide to assign it to somebody else. What I always like to say is that songwriting is an art, not a business. There is no money generated by the act of writing a song; it's just what you do. Music publishing is the business of turning songs into something that earns money. So if you want to earn money from your songs, you have no choice but to take on that role of the music publisher. Songwriting is the act of creating the music, and publishing is looking at your creation and thinking of revenue outlets for it.*[1]

Eric Beall, author of *Making Music Make Money, An Insider's Guide To Becoming Your Own Music Publisher* and the Berkleemusic online course *Music Publishing 101*, and former VP of Creative for Sony/ATV Music

I couldn't have said it better and I totally concur. Of course, if you really want to help yourself, start with using some basic business tools to assist you in discovering what you're good at. Then ask someone else to analyze your music and songs, someone who you respect and trust. Use the following framework to help you to do this.

---

[1] http://www.artistshousemusic.org/articles/music+publishing+101

The Art of Music Publishing. DOI: 10.1016/B978-0-240-52235-7.10008-9

THE UNDISCOVERED COUNTRY

It's so hard to analyze yourself or something close to you. The following framework allows you to try and become objective, a third party, and look in on what you are doing.

What do you consider are your best songs to date? What has brought you to that decision? What do you particularly like about the songs you've selected, and how have third parties commented on them?

Focus on such things as structure, lyrics, melody, chorus, bridge, length, hooks, start, ending, recording, vocals, instrumentation, presentation, key, genre and style.

You could analyze each song to help uncover your skills; this might assist you in looking at areas you can improve. Perform a SWOT analysis, looking at your strengths, weaknesses, opportunities and threats.

Being self-critical can be very hard and so it's wise to speak to people you know and trust for some balanced critique of your work.

Your analysis might highlight some specifics that could assist you: perhaps the verse is strong but takes too long to get to the chorus. Or there's no real lift into the chorus, or the song needs a new bridge. Or the lyrics are strong and well crafted but not relevant to this genre of music or customer. Perhaps the demo of the song is poor as the vocals are out of tune and therefore the song is not well represented.

- **Strengths** – Good ability to write lyrics, but they need better focus. Your song has a strong verse and a chorus you can't forget, as everyone goes around humming it.
- **Weaknesses** – Potential issues with structure, bridge to chorus, recording and presentation.
- **Opportunities** – The need to explore the development of songwriting structure could lead to developing a strategy to co-write with other writers who may complement your weaknesses, thus turning a weakness into a strength. How might you find such people (publishers, managers)? What networking events might you meet them at? Can you make appointments to go and see people? How can you develop your writing skills (courses, learn from others)? What composers do you admire? Studying their work may provide an insight into structure and the other elements listed in Weaknesses. More opportunities to perform live are needed to gain audience feedback.
- **Threats** – If you don't focus in on your weaknesses then you won't improve your craft of songwriting. There are plenty of other composers striving for what you want and therefore to be heard above the competition your songs and compositions need to be well crafted and well received. Your career won't progress as quickly as you want if you don't realize that all skills, all talents can be developed. A sensible, business-like approach is needed, along with dedication and focus and allowing your art to breathe.

The purpose of using this type of framework is that it helps you to unravel different elements. In doing so it acts as a signpost and then through further research and work you might find a way to turn weaknesses into strengths and move your career forward.

This is a very simplistic solution and of course there are lots of holes in it. Beauty can be in the eye of the beholder. How many composers, artists and bands get rejected but go on to do great things? In truth, very few! Most of the good ones get spotted eventually, if not immediately.

External pressures also come to bear on decision making, within the world of marketing, consumer behavior and business economics (i.e. profit).

Perhaps before you analyze yourself, analyze other writers and other songs and think about structure, content, lyrics, and so on. Practice the art of publishing A&R.

However, back to the key topic: how to get a publishing deal.

Once you feel you are ready and your songs are ready, how and where do you start?

- **Network**. Singer—songwriter nights are often sponsored or put on by your local collection society. Also find out about other industry networking events.
- **Join your local collection society.**
- **Speak to lawyers** and get advice on who to talk to in the industry.
- **Read the music press** and pick up on publishing company activities and staff.
- **Contact publishers** (major, major independents and independents). Where do you get lists of companies? The *Music Week* directory is a good place to start; the Music Publishers Association will have a list of members. The same applies in the USA: BMI and SACAP have great contacts and A&R worldwide.
- **Research relevant companies**. Many have websites. Find out who they represent and what success they have. Check these companies with a lawyer for advice and input as to who would be good to go and see.
- **Ring and make an appointment**. Sending CDs in can be quite futile and I wouldn't recommend this unless you call ahead, you have been recommended or you have already met the person and they asked you to send something in. In the digital age an MP3 (1 or 2) is fine, but never send any more than one or two tracks. If you do it's usually a sign that you have no idea which songs are good.

PRODUCT AND PRESENTATION — SONGWRITERS

Ask yourself: who are my customers? They may be artists looking for songs, or managers managing artists who are looking for songs. Labels sign artists, many of whom don't write their own songs, and so the label looks for songs. Music publishers need songs for these artists, so they need writers. Songwriters have various routes to success (routes to market).

If a songwriter is great at recording material and producing music as well as writing it then other opportunities open up. These include composing and recording music for television commercials (nearly every commercial you see on television has music synchronized to it; some of the music is well known, but a lot is newly composed).

If you are a songwriter and a record producer then this opens up yet another world. In your role as a record producer you will get to work with lots of labels, artists and managers, and the opportunity to co-write with the artist increases. Publishers like to work with songwriters who are also producers or artists, as both increase the likelihood of songs being commercially activated.

As an example, my husband has co-written and produced artists such as Beverley Knight, Liberty X, Ultravox and Jaki Graham, and as a result the songs were recorded and released by the artists' respective labels all over the world. Royalties for the songwriter and music publisher are then created and collected accordingly.

What will a publisher be looking for? Again, reading the earlier chapters will have given you a better perspective. Essentially they are looking for songs that will create, or are currently creating, an income stream. The music industry is a business and so too is music publishing. If the songs you've written have had no exposure, have not been recorded properly, and there's no artist interested in using them, and in addition the song structure is weak and there is no obvious way of creating income from it, then you will be wasting your time and that of the publisher even approaching anyone for representation.

If you make a concerted push into the industry before you are ready you may risk burning a few bridges along the way. So it's important to hold back until you've got some songs together and you have genuine feedback and reaction to them is strong (and not from your mum or your girlfriend/boyfriend). The more you can do to self-promote yourself and get some good-quality demos done the better. The investment in equipment needed to achieve this is relatively low compared with 10 years ago. A good computer (e.g. iMac) and the right software and you're up and running.[2] If you can generate some income yourself or generate some strong interest in your songs before seeing music publishers it only strengthens your negotiating position.

SUMMARY

- Don't go to market too quickly.
- Make sure your songs are well structured, with good lyrics and a great chorus. Make sure the chorus hits home within the first minute of the song. Work at it! You don't wake up one day a brilliant songwriter and musician. Put the work in and you'll have a better chance of success.
- Make sure the demo is of good quality.
- Get some proven feedback that supports the strength of the songs (website, MySpace or similar).
- Be self-critical and use a SWOT analysis along with feedback from those you respect to help you improve.
- Try and create your own outlets for your songs.

---

[2]www.planetaudiosystems.co.uk

- Remember, in the early days you are both the songwriter and the publisher, so put your business head on and think about how you can get your music heard. The more you can do yourself the better.
- Register with your local rights society — PRS for Music, BMI, ASCAP; all international societies detailed in earlier chapters (see Chapter 3 for the minimum criteria to fulfill).
- Get out and get networking — sitting in your bedroom and just going online doesn't count.

UK and international trade shows

9

In the music and creative industries we have trade shows that take place in the UK and all over the world. As a music publisher, there is a broader base of trade shows to be considered which expand into the television advertising arena, film industry, television industry and computer games shows. Remember that each of those business sectors has a substantial appetite for music. Let's take a look at some of the shows that you should consider.

BENEFITS OF LOCAL AND INTERNATIONAL TRADE SHOWS

Trade shows have many benefits:

- They are cost-effective — many countries under one roof.
- They cover a broad base of companies — all genres.
- They have international representation.
- They are a good place for international networking.
- You can get to know local territories.
- You can secure individual deals (single market) or international deals (many markets).
- Often trade show grants are available from local government offices to assist start-up companies to gain a foothold in the export market.
- They provide an excellent opportunity:
  - to hear international music — attending parties, shows, artist presentations
  - to network with people from your own country
  - to find international representation — lawyers, accountants, agents, live work, deal shoppers, distribution, labels and publishers all under one roof
  - to attend seminars which assist your training and understanding of new markets and industry issues.

EXAMPLES OF TRADE SHOWS
MUSEXPO[1]

Their website is packed full of excellent detail. They always have a high-caliber list of industry speakers in attendance. This is an international trade show taking place several times throughout the year.

---

[1]http://www.musexpo.net

The Art of Music Publishing. DOI: 10.1016/B978-0-240-52235-7.10009-0

Following six successful years of the event based in West Hollywood, California, MUSEXPO has once again taken its conference global. MUSEXPO Europe took place in June 2010 in the heart of London, England at The Cumberland Hotel yards away from London's Marble Arch and other historic landmarks.

The format of the conference, as well as the hosted executive networking breakfasts, lunches, tea breaks and cocktail sessions remained at the heart of the event, followed each evening by MUSEXPO's infamous live showcases by up-and-coming artists, taking place in London's premiere venues.

The event has attracted hundreds of the most influential music, media and technology executives from 30+ countries across four continents.

I have attended MUSEXPO on several occasions. It is aimed more at the top end of the networking sphere and those who are ready to showcase artists to a select audience, and those who can consider working with such artists. I have always come back with a rich list of networks and have found their conference program quite exceptional, with great insight into new trading territories and routes to market.

The Great Escape[2]

This is very much a conference growing in stature.

It considers itself "Europe's leading festival for New Music". It is divided into segments of conference, with a broad nightlife with venues and gigs from new and emerging talent. It is a celebration of music through and through. Conference speakers are diverse, from managers through label executives to economists. The conference is well received by those coming into the industry and moving through the industry. It is a great networking event based in Brighton, England.

In The City[3]

A long-running and quite historic music conference based in Manchester, England. Its popularity has come and gone and come back again. It's all about the music, going out to the clubs and music venues and discovering talent. There is a networking and conference panel each day, which is well attended. It is supported more by the UK industry than by international customers.

MIDEM[4]

The granddaddy of conferences, it combines many of the above elements but its focus is international networking and buying and selling (licensing and contracting) for businesses all over the globe. It's the professional face of the industry and more of

---

[2]http://www.escapegreat.com
[3]http://www.inthecity.co.uk
[4]http://www.midem.com

a business-to-business conference. It is a place where deals can be done, networks forged and long-lasting deals developed. I'm a big fan ... I've concluded or seedbedded hundreds of thousands of pounds worth of business at MIDEM over the years: everything from setting up subpublishing or administration deals, through licensing product for synchronization use to licensing product to major and independent labels. In 2010, I took 20 new business start-ups to MIDEM and through the MTV listening sessions secured a contract with MTV for use of their music for upcoming episodes of *The Hills*. Joe Cuello, VP of Creative Music Integration at MTV, was seriously impressed with their music. This sort of early success is invaluable. The band to watch out for is Arcady Bliss (http://www.myspace.com/arcadybliss).

Popkomm[5]

Popkomm is making a comeback in 2010 with new dates, strong partners and THE new location. Popkomm 2010 — consisting of marketplace, festival, conference and a BRAND NEW B2C area — will be an integral part of the new umbrella brand 'Berlin Music Week', and will take place 6—12 September. The event will introduce all facets of the capital's music industry and culture to international visitors and music business professionals.

Popkomm 2010 will land at THE extraordinary location — Airport Berlin Tempelhof. The international trade show will take place in this very historical building, often referred to as the 'mother of all airports', and still today's third largest integrated building in the world (after Peking airport and the Pentagon). Popkomm will provide several networking possibilities and meeting options within its market place. Popkomm will partner with the festival makers of the Berlin Festival which will take place for the sixth time in the hangars and on the airfield of the Airport Berlin Tempelhof on the 10th and 11th September and expects more than 15,000 music lovers to attend daily.

Exclusive showcase events will take place at the Popkomm Festival during the daytime. In charge is Paul Cheetham, who has almost 20 years of experience as a music promoter, artist and event manager, concert producer and music industry consultant. Check for latest information about the festival at http:// popkommlive.com.

I have attended on and off since 1990, and having a European focus to your business and product is superb.

Cannes Lions[6]

The Cannes Lions International Advertising Festival is the world's only truly global meeting place for professionals in the communications industry. It offers

---

[5]www.popkomm.de
[6]www.canneslions.com

seven days of award ceremonies, seminars, workshops, master classes, exhibitions, screenings and networking.

It costs a lot of money to attend but once your music publishing catalogue is generating income perhaps this event may break down the doors in the advertising arena.

Winter Music Conference[7]

Winter Music Conference, in its 25th consecutive year, is one of the most publicised annual music gatherings in the world. A pivotal platform for advancement of the industry, WMC 2009 attracted 1910 artists and DJs, 3228 industry delegates from 62 countries and over 70,000 event attendees for a concentrated schedule of more than 500 events presented across five days. Music, as one of the world's most accessible cultural art forms, gives WMC the unique ability to cross economic, geographic and social boundaries. Over 1.3 million visitors from 183 countries log on to the WMC website each year.

The WMC is particularly good for dance music companies, producers, remixers, and meeting radio and club DJs.

SUMMARY

I've been lucky enough to have attended all these trade shows and even some like The New Music Seminar, held in New York in the 1990s, that are no longer operating. With clear aims and objectives these trade shows can be the gateway to achieving new networks, new business and an international sales base. They provide a meeting place to make new contacts, stay in touch with existing networks, build and develop ongoing relationships. Trade shows can speed up business integration in the global economy. While they are expensive, they are a cost-effective way of achieving international success. Each year I take up to 20 business start-ups to such events to assist them in getting their business ideas off the ground or in growing their networks for international business. I'm always very impressed with what they achieve.

One of our family businesses has taken conferencing to another level. We are currently breaking the doors down with a worldwide tour, setting up distributors for a new product we've developed and new products ongoing with patent applications pending. UK Trade & Investment (UKTI) has been exceptionally helpful and is available for all export initiatives for UK business.[8] You might like to take a peek at www.rotolight.com

---

[7]www.wmcon.com
[8]www.uktradeinvest.gov.uk

Networking — building a sustainable network

10

WHAT IS NETWORKING?

It is the most important marketing tactic in developing and sustaining relationships with individuals and organizations from which you can receive assistance, and provide assistance to.

Networking is a two-way street. The better you are at it, the further you will go in business. Your network can be a catalyst for opportunities, ideas and business, on an immediate or a future basis. So often, though, people think it's about collecting business cards and stripping out all you can in a raiding party from a given meeting or encounter. This is not networking. Networking is first being open and willing to share. It moves on to looking at how both parties can benefit from the relationship, maybe not immediately (and you may be able to assist them before they can assist you), but at some point giving back some genuine business so that both parties benefit over time. It's about trust, recommendation and friendship, which have to be earned for networking to be sustained and have longevity. I have some lovely business friends with whom I meet every few months and we have no set agenda, just to meet and catch up, and in doing so we hope that there may be opportunities in business that may assist the other person. Some people are very guarded (I am not one of them), some people think they can only get ahead if they keep their contacts private (I am not one of them), and some people feel it is a strategic object to build a network that only they will communicate with. Isn't that actually the opposite of networking? Not only do I enjoy my work but I enjoy seeing others, and my friends, business associates and students flourish and become successful. That's how I network, I facilitate introductions where they can make a difference, and whenever I need help or guidance I go to my network and seek it out. My lawyer, Alexis Grower, has for the past 20 years provided me unselfishly with contacts and introductions in case they should help my business and work, and I have always tried to return the gesture. Think about networking and what it will mean to you.

Perhaps you're new to the business or want to build contacts in music publishing. Start by identifying ways to build your network. Where can you network, how will networking take place (formal or informal) and which industry organizations do you want to join? Before you attend any functions research the key people who will be attending. Identify key people of influence, what they do and what you may have to offer them. Ignorance is not very pretty. You wouldn't go to an interview without

The Art of Music Publishing. DOI: 10.1016/B978-0-240-52235-7.10010-7

knowing all you can about the company you might like to work for, would you? So don't go to an event without knowing something about the people hosting the event. Start by introducing yourself to them on arrival; be polite and confident. I very much enjoy networking and helping to facilitate business for others. I love meeting and finding out about new businesses: it's a great buzz.

I am fortunate in having a great network that spans many years, many countries and many business sectors, but mainly within music and media. I care about my network; I help as many people as often as I can and in return whenever I need to call on my network they are always there. This book is a testament to that.

SOME USEFUL NETWORKING TIPS

- Always have some business cards with you.
- When receiving a card, make a note on the back about who gave you the card, the conversation you had, and any key words to jog your memory the next day.
- Try to visualize their name and face and something to do with their image. Remembering people's names is a great skill. I'm not the best, I have to admit, and that's why I try to do a lot of image association with a name and business card.
- Send an e-mail after the event to remind them that you met, what you discussed and any other follow-ups that are relevant.
- Keep the business card in a directory and make time to check visually every once in a while in case something comes up that they may be perfect for.

SUMMARY

Networking is one of the most important skills to learn for everyone: the music publisher, the business owner, the songwriter and the manager. Communicating with other people is a skill that you will get better at the more you do it. The wonderful thing about networking is that it's face to face. It gets you out of the office, or out of the studio, and puts you at the 'coal face'.

CHAPTER

The producer and the song 11

Rod Gammons (producer/writer)

It would be hard to write about the role of the record producer without first highlighting the extraordinary work of the early pioneers of modern record production.
The first sea-change happened in the 1950s, with the birth of rock 'n' roll when
producer Sam Phillips opened his Memphis Recording Services studio, signing Elvis
and creating Sun Records. However, what we now consider to be modern record
production started following a second dramatic change in musical style and
recording techniques that kicked off in 1962 with the song 'Telstar', produced by Joe
Meek, closely followed by releases from The Beatles produced (and orchestrally
arranged) by George Martin, the mighty 'wall of sound' recordings by Phil Spector,
and the imaginative Brian Wilson (The Beach Boys). These legendary producers
redefined the way that records would be made forever, leaving a lasting legacy of
imaginative excellence and setting the benchmark for all that followed.

Music is the only product that is consumed before it is purchased.

Quincy Jones

Those are the words of the greatest friend a song ever had, Quincy Jones (27
Grammy Award winner and the producer of Michael Jackson's 'Thriller' and 'Off
The Wall' albums, and composer of 33 movie soundtracks, whose works have sold in
excess of 200 million albums worldwide). His point was that someone would hear
a song (apparently for free) 10 times a day on radio and perhaps even hear it again on
television or in a film soundtrack, but the consumer would still be driven to go out
and purchase a copy to own for themselves, such being the power of music!

From a producer's perspective, it is important to understand the relationship
between the artist, the recording and the song. I like to use a simple analogy, which is
that the artist is like a fabulous Formula One racing car, the song is the fuel and the
recording is the way that it is driven. Imagine the most beautiful racing car in the
world with the wrong fuel, driven well … it's not going to get very far. Alternatively,
an average car, fueled by rocket fuel, driven well is going to go a long, long way.

Nothing is as important as a good song: it all starts with a song, and in the end the
longevity, career distance and staying power of an artist are all decided by the
strength and quality of the songs they record. A hit song is 'the gift that keeps on
giving'.

Quincy Jones knew the importance of choosing the best songs, and he would task
his artists to make the very best judgments about the songs to go on their album.
David 'Hawk' Wolinski, the keyboard player of Rufus and Chaka Khan, explained

The Art of Music Publishing. DOI: 10.1016/B978-0-240-52235-7.10011-9

Quincy's technique: "Every member of the band was each given a large box full of song demo tapes to listen to, and these songs had been solicited from both the usual industry sources as well as the unusual. Each band member was then asked to choose the best three of these 'third party' song demos and these songs would be added to the catalogue of songs the band had written themselves". Finally, a band A&R meeting would be convened, and the final selection of the songs to be recorded would take place. This way the final 10 songs to be recorded would have been selected from literally thousands of songs, ensuring the very best 'fuel' for the artist's next album. They would record one or two more tracks than they intended to release, so the final A&R process could refine the album running order by dropping any recordings that did not work out as well as intended.

Sometimes, this refining process does not always go as intended, and Hawk recounted the extraordinary tale of his song 'Ain't Nobody'. When Rufus finished recording their 'Stomping At The Savoy' album they presented the recording to their A&R team at Warner Brothers. Hawk was astounded when the A&R guys said they didn't see it as a single, so he threatened to pull it from the album, telling them that Quincy wanted it for Michael (Jackson), who was recording the 'Thriller' album. Reluctantly, the A&R guys agreed to release it as a single and the rest, as they say, is history: it was a number one R&B hit around the world, and won a Grammy for best R&B performance. In hindsight, Hawk was really grateful that Michael didn't record it, because not many of Michael Jackson's hits have been successfully covered by other artists. To this date 'Ain't Nobody' has been re-recorded by literally hundreds of different artists around the world (most notably LL Cool J, Jaki Graham, KT Tunstall, Gwen Dickie/KWS, Liberty X, Daniel and Natasha Bedingfield, The Course, George Michael, Diana King, Kelly Price, and many others) and the song has been a No. 1 hit somewhere in the world in every decade following the original success with Rufus.

What is the role of a record producer? He is expected to be all things to all men, sometimes doing very little, sometimes doing almost everything (A&Ring, writing, performing, arranging, recording, engineering, mixing, mentoring, nurturing, and guru), and the best record producers achieve this brilliantly. Hawk once commented that Quincy seemed to spend most of his time sitting at the back of the studio reading the *Wall Street Journal*, but all of the brilliant work had already been done outside the studio as he had assembled an inspired team of the engineer Bruce Swedien and musicians Bill Reichenbach, Gary Grant, Jerry Hey, John Robinson, Hawk Wolinski, Stevie Wonder, Louis Johnson, David Foster, Steve Lukather, Greg Phillinganes and the rest of the usual suspects. Typically, these outstanding musicians would instinctively know what to contribute to the recording, and Quincy could carry on checking his stock prices. Artists and bands can have a strong sense of audio identity, but can exceed their own expectations when teamed up with the right producer, whose experience, encouragement and technical expertise can bring out the very best in both the artist and the song. Vocalists often forget that they can't be both sides of the mic at the same time, i.e. singing and listening, and getting the right vocal performance on the recording is vital. A good producer will win the trust of the artist, and become the interface between the artist and the label (sometimes this can be a very difficult path to travel); sometimes the song will require some rearranging, or even some rewriting, and if this is going to happen then it's best to agree the terms up front on which those contributions to the song take place, or risk losing your share of the copyright. An arranger's share is typically 1/12th of the work; if you contribute lyrics or music then that share should be negotiated and agreed, remembering that music and lyrics represent 50% each of the song as a whole. Quite often the producer may also be the writer/co-writer/publisher of the song, in which case the record label may impose a controlled composition clause on the payment of mechanical royalties for the album. This is very common in the USA and many other countries where a statutory mechanical royalty rate applies.

Todd Brabec and Jeff Brabec state a few interesting points on the ASCAP website which I will share with you here.

> It should be mentioned that the per-song statutory mechanical royalty could be reduced under certain circumstances. However, such reduced rates are voluntary and occur only if the publisher agrees or if the songwriter is a recording artist and has to accept such lower royalties in the record company contract.

> Many agreements, the majority, in fact, contain language which provides that if the recording artist or producer has written or co-written a song, has ownership or control of a song, or has any interest in any composition on the album or single, the mechanical royalty rate payable by the record company for that composition is reduced. Such compositions are referred to as controlled compositions. Most contracts attempt to establish a 75% rate (specifically, 6.82¢, which is three-quarters of the 9.1¢ full statutory rate) for all controlled compositions.

Note here that success can influence change and an artist/writer can insist that full statutory mechanical rate is paid and certainly your lawyer should try to limit any contractual affect. The position of record labels in this regard dates back many years and you would expect in this day and age that such matters would not continue, but they do, so beware.

> In other cases, the record company will establish a maximum aggregate mechanical penny royalty limit for an album (for example, 10 songs × 6.82¢ = 68.2¢ per album). In a sense, a cap on royalties. Under these clauses, the artist or producer guarantees that he/she will secure reduced mechanical rates on all songs on the album so that the maximum penny rate (e.g. 68.2¢) payable by the record company to music publishers and songwriters for all songs is not exceeded. If this maximum aggregate album-royalty rate is surpassed — for example, if the writer/artist wants to put 12 songs rather than 10 on the album — the difference is normally deducted from the artist's or producer's record, songwriter and publishing royalties, or, the per-song royalty rates for the writer/artist or writer/producer will be reduced proportionately.

Let's take a look at how this arithmetic affects a specific situation: let's say that the writer/performer has a 10-song × 6.82¢ maximum royalty rate on his/her album (in other words, 68.2¢ total) and, instead of writing all 10 songs, writes only eight and records two songs from outside writers who demand the 9.1¢ statutory 2007 per song. In this case, the mechanical royalties would look like this:

68.2¢ Album-royalty maximum payable by record company
−18.2¢ Two outside songs at 9.1¢ each =
50¢
/8 The number of artist-written songs =
6.25¢ Per-song royalty to artist/writer and publisher

As you can see, the writer/artist's mechanical royalty has been reduced to 6.25¢ per song from 6.82¢ per song due to the inclusion of two outside-written songs on the album. By the same token, as the writer/artist records more outside-written songs, the artist's per-song royalties for his/her own works will be further reduced. Sometimes, in fact, when a writer/artist has recorded a substantial number of songs by other writers, he/she has been put in a position of receiving no royalties for his/her own songs, since the aggregate album-royalty maximum has been paid out to outside songwriters and publishers. Ouch! But it can get even worse. There have actually been instances in which the writer/ artist's mechanical royalties have been in the minus column for every album sold because of the operation of these controlled-composition clauses. Additionally, the era of multiple remixes has given rise to a clause which provides that the writer/artist will only receive a mechanical royalty for one use of his/her song regardless of the number of versions contained on the single or album.[1]

Todd and Jeff's book is a great read and provides a very coherent picture especially of the American market. The UK market (I am glad to say) is not as onerous. The artist and producer still suffer reductions and deductions of their own royalty but this does not affect the mechanical royalty rate applied.

Developing every ounce of commercial potential out of a recorded copyright is really the primary role of the producer. Great producers get the best out of the arrangement, recording and performance of the song; exceptional producers exceed this expectation with inspired voyages of imagination. One such producer is Trevor Horn: three times winner of the Brits' 'Best Producer' award, Grammy Award winner for record of the year (Seal's 'A Kiss From A Rose') and an illustrious songwriter, musician, artist and later producer, Trevor went on to form his own publishing company, Perfect Songs, in 1982, with his wife, Jill Sinclair.

---

[1] © 2007 Todd Brabec and Jeff Brabec; As taken from ASCAP's website but with reference to *Music, Money and Success: The Insider's Guide to Making Money in the Music Business* (Schirmer Trade Books/Music Sales 502 pages) www.musicandmoney.com

Trevor Horn's songwriting credits date back to 1979 when he co-wrote a song for Dusty Springfield, 'Baby Blue', with Bruce Woolley and Geoff Downes. All the Buggles' hits — including 'Video Killed The Radio Star', 'Living In The Plastic Age', 'Elstree' and 'I Am A Camera' — were co-written by Horn and Downes and, occasionally, Bruce Woolley. Horn also co-wrote all of the 1980 Yes album, 'Drama'. On his return to the band (as producer) in 1983 he contributed to their biggest hit, 'Owner Of A Lonely Heart' and the dance hit 'Leave It'.

During 1982 and 1983, Horn worked with Malcolm McLaren and Anne Dudley, writing numerous worldwide hits including 'Buffalo Gals', 'Double Dutch', 'Duck For The Oyster' and the 'Duck Rock' album of world beats and new hip-hop styles.

In 1984, he co-wrote several classic hits with the Art of Noise including 'Close (To the Edit)', 'Beat Box' and 'Moments In Love'. The next year he co-wrote 'Slave To The Rhythm'. This was originally intended as Frankie Goes to Hollywood's second single, but was instead given to Grace Jones. Horn and his studio team reworked and reinterpreted it, jazz style, into six separate songs to form the album 'Slave To The Rhythm'.[2]

It was widely reported at the time that part of the inspiration behind the 'Slave To The Rhythm' project was that Trevor Horn had only been commissioned to record one song with Grace, so Horn worked around that limitation by rearranging, readapting and remixing the same 'song' many times to come up with a complete album, based around a single song copyright.

Released in 1985, 'Slave To The Rhythm' is a unique conceptual mini album from influential artist Grace Jones. Using revolutionary production techniques, Horn enhances Jones' performance resulting in an album, which embraces art and fashion with extraordinary results. With only interpretations of the title track on the record, it includes special 'appearances' from Ian McShane, John-Paul Goude and Paul Morley.

'Slave to the Rhythm' was written by Bruce Woolley, Simon Darlow, Stephen Lipson and Trevor Horn. The album reached No. 12 on the UK Album Chart in November 1985 and No. 10 on the German Album Charts. The single became one of her greatest commercial successes and counts as one of Jones' signature tunes.

On the 'Slave To The Rhythm' album the hit single version of the track is confusingly retitled 'Ladies and Gentlemen: Miss Grace Jones'. The track called 'Slave To The Rhythm' on the album is in turn an entirely different interpretation of the song — a well-known fact that somehow managed to elude the producers of a recent hits compilation — which led to record company Universal Music issuing the wrong version of one of Jones' greatest hits on one of their many best of packages, see 'The Ultimate Collection'.

---

[2] http://en.wikipedia.org/wiki/Slave to the Rhythm

Horn's songwriting can be heard on numerous film soundtracks. In 1992, Horn collaborated with composer Hans Zimmer to produce the score for the movie Toys *starring Robin Williams, which included interpretations by Tori Amos, Pat Metheny and Thomas Dolby.*

In the 2000s, Horn provided additional production on three international hits for t.A.T.u., 'All The Things She Said', 'Not Gonna Get Us', and 'Clowns (Can You See Me Now)'. He also wrote 'Pass The Flame' (the official 2004 Olympic song) and co-wrote the title track from Lisa Stansfield's 2004 album 'The Moment'.[3]

SAMPLING, LOOPS AND SAMPLERS

Since the introduction of sampling technology there has been a great temptation to 'repurpose' small audio extracts of previous artist's recordings (i.e. loops). While the assumption is that there is a general rule that you can't copyright a short except of a drum beat, this is actually not the case, and this area of modern music production is probably the one that causes the most problems.

On a recording there are two copyrights, one in the recording (i.e. ownership of the recording) and one in the music (i.e. the song or musical work that has been recorded). For a sample to be used in recording there has to be consent from both parties, and this is usually the subject of negotiation with both the record company that owns the recorded sample and the music publisher of the song. The duty of keeping track of possible samples and excerpts of musical copyrights used within the recording process falls on the shoulders of the record producer. This obligation is usually strenuously dictated in the producer's contact, and failure to observe/control this aspect of the recording process usually ends in expensive deductions of royalties or, in the worst case, legal proceedings and recovery of damages with costs. It is therefore a very good idea for the producer to keep and fill in a sample clearance form on every session, which is passed back through to the artist's record company A&R coordinator, who will endeavor to negotiate with the third party copyright owners before the recording is finished, so the sample could be removed if no agreement can be reached.

It is always worth considering the alternatives to using samples:

- Write something similar (it only needs to be a few notes/nuances different) and record a replacement sample so that both the recording and the musical copyright are owned by the artist, songwriter, record company and publisher.
- Negotiate with the sample's publishers and record company, ahead of the recording process, and try to get a licence where a small one-off fee is paid instead of a royalty: it will be much cheaper in the long run.

[3]http://en.wikipedia.org/wiki/Trevor Horn

HOW TO GET IT HORRIBLY WRONG – VANILLA ICE, 'ICE ICE BABY'

The classic example of how not to do it is Vanilla Ice, and his hit record 'Ice Ice Baby'.

Robert Matthew Van Winkle (born October 31, 1967), best known by his stage name Vanilla Ice, is an American rapper. Born in Dallas, Texas, and raised in Texas and South Florida, Van Winkle released his debut album, 'Hooked', in 1989 through Ichiban Records, before signing a contract with SBK Records, which released a reformatted version of the album under the title 'To The Extreme'. Van Winkle's single 'Ice Ice Baby' was the first hip-hop single to top the Billboard charts. 'To The Extreme' became the fastest selling hip-hop album of all time, peaking at No. 1 on the Billboard 200. The album spent 16 weeks on the charts, and sold 11 million copies.

No attempt was made to clear the sample or clear the use of the sample or music copyright prior to the release of 'Ice Ice Baby', and during the first couple of MTV interviews during the record's release Van Winkle hilariously claimed that he 'had not' sampled Queen and David Bowie's hit single 'Under Pressure'. As soon as the record became a hit, there followed a writ, and the royalties were frozen.

The song 'Ice Ice Baby' was initially released as the B-side to Van Winkle's cover of 'Play That Funky Music', but the single was not initially successful. When a disc jockey played 'Ice Ice Baby' instead, it began to gain success. 'Ice Ice Baby' was the first hip-hop single to top the Billboard charts. Topping the Australian, Dutch, Irish, Italian and UK charts, the song helped diversify hip hop by introducing it to a mainstream audience.[4]

[4]http://en.wikipedia.org/wiki/Vanilla ice Source: Wikipedia.

HOW TO GET IT HORRIBLY WRONG (AGAIN) – THE VERVE, 'BITTER SWEET SYMPHONY'

The following is an extraordinary example of what can go wrong if you don't get all the business sorted out before the recording gets released. The court case surrounding this song is based not on the song or any part of the song that was written by the original composers, but on a sample of an arranger's recording of the song, in which the arranger added a riff played by strings, an entirely new musical phrase, and the use of this 'arranger's addition' in another musical work prompted the awarding of 100% of the songwriters' royalties to Keith Richards and Mick Jagger instead of The Verve.

Bitter Sweet Symphony' is a song performed by English alternative rock band The Verve, and includes music written by Keith Richards and Mick Jagger of The Rolling Stones. It is the lead track on The Verve's third album 'Urban Hymns' (1997). It was released on 16 June 1997 as the first single from the album, reaching No. 2 on the UK Singles Chart. The song's momentum built slowly in the United States throughout the latter months of 1997, ultimately leading to a CD single release on 3 March 1998, helping the song to reach No. 12 on the Billboard Hot 100. The song also became famous for the legal controversy surrounding plagiarism charges.

Rolling Stone ranked 'Bitter Sweet Symphony' as the 382nd best song of all time. In May 2007, NME magazine placed 'Bitter Sweet Symphony' at No. 18 in its list of the '50 Greatest Indie Anthems Ever'. In September 2007, Q published a list of 'Top 10

Tracks' as selected via a poll of 50 songwriters; 'Bitter Sweet Symphony' is included. In the Triple J Hottest 100 of All Time, 2009 (an online music poll conducted by the Australian radio station Triple J) the track was voted the 14th best song of all time. Although the song's lyrics were written by Verve vocalist Richard Ashcroft, it has been credited to Keith Richards and Mick Jagger after charges by the original copyright owners that the song was plagiarized from the Andrew Oldham Orchestra recording of The Rolling Stones' 1965 song 'The Last Time'.

Originally, The Verve had negotiated a licence to use a sample from the Oldham recording, but it was successfully argued that The Verve had used "too much" of the sample. Despite having original lyrics, the music of 'Bitter Sweet Symphony' is partially based on the Oldham track, which led to a lawsuit with ABKCO Records, Allen Klein's company that owns the rights to the Rolling Stones material of the 1960s. The matter was eventually settled, with copyright of the song reverting to ABKCO and songwriting credits to Jagger and Richards.

"We were told it was going to be a 50/50 split, and then they saw how well the record was doing", says band member Simon Jones. "They rung up and said, 'We want 100 per cent or take it out of the shops, you don't have much choice'."

After losing the composer credits to the song, Richard Ashcroft commented, "This is the best song Jagger and Richards have written in 20 years", saying it was their biggest hit since 'Brown Sugar'. The song was later used, against the will of the band, by Nike in a shoe commercial. As a result, it was on the 'Illegal Art' CD from the magazine Stay Free! *The song was also used in a Vauxhall Motors advertisement and several of Opel, prompting Ashcroft to declare onstage at their homecoming performance at Haigh Hall, Wigan, in May 1998, "Don't buy Vauxhall cars, they're shit". However, the band were able to stop further use of the song by employing the European legal concept of moral rights.*

On Ashcroft's return to touring, the song traditionally ended the set list. Ashcroft also reworked the single for 'VH2 Live' for the music channel VH1, stripping the song of its strings. Ashcroft is quoted as saying during the show: "Despite all the legal angles and the bullshit, strip down to the chords and the lyrics and the melody and you realize there is such a good song there".

He also dedicated the song to Mick Jagger and Keith Richards during a gig at the Sage Gateshead in Gateshead. After several audience members booed, Ashcroft exclaimed, "Don't boo, man. As long as I can play this song I'm happy to pay a few of those guys' bills". In a Cash For Questions interview with Q *magazine published in January 1999, Keith Richards was asked if he thought it was harsh taking all The Verve's royalties from 'Bitter Sweet Symphony', to which he replied, "I'm out of whack here, this is serious lawyer shit. If The Verve can write a better song, they can keep the money".[5]*

While it is clear that the record producer is not necessarily in total control of the building blocks of the artist's compositions, the producer can play an important part in advising the artist during the recording process, perhaps by suggesting freshly recorded alternatives to the problematic sample parts or assisting in 'rewriting' the sample. In the case of The Verve, it would have been very easy to pastiche the Andrew Oldham riff, changing a few notes here and there to come up with something functionally identical which retained the melodic style, structure and meter, then re-record the part with a small string section and save the royalties (and future royalty income) for the artist/writer. When the Stones threatened to get the record pulled from the shops, this process could have been completed in less than eight hours (total recall on the mix, re-record the strings, mix and master) and fresh stock could have been inside the shops within 24 hours. If the producer undertakes this work as an arranger, then he should be entitled to an 'arranger's share' of the royalties from that version of the work, normally 1/12th of the royalty income. I would be very surprised if the Rolling Stones actually paid this arranger's share to Andrew Oldham, either on his recording of the work or on the subsequent Verve adaptation of Oldham's riff, but I guess they can afford better lawyers.

[5]http://en.wikipedia.org/wiki/Bitter Sweet Symphony

TECHNOCLASSIX

An interesting example of working within the rules is the 'Technoclassix' album. In this album I licensed excerpts of very recognizable classical works from Beethoven, Bach and Mozart from a leading Classical Catalogue label (Conifer Records) (the only rule was the composers had to have been dead for more than 75 years so their works were 'public domain'). I then commissioned leading dance remixers to adapt those recordings and turn them into techno tracks. We sent the recordings to PRS so their copyright panel could assess the arranger's share (putting the most extreme adaptations first on the CD!) and the panel subsequently awarded 12/12ths (i.e. 100%) of the composer's royalty to our remixer/arrangers. The album was then published through our publishing company. We had a lot of success licensing the 'Technoclassix' album in most major territories where there is an active dance music scene. Once a copyright work becomes public domain it can be rearranged and this arrangement can be published. This is how much of the printed sheet music publishing industry works, publishing new arrangements of popular classical works by famous composers.

One excellent business model for entrepreneurial producer/composers is that of a 'production stable', where the producer assembles a team of like-minded writers, engineers, programmers and remixers who are managed and published by his company. Usually the producer will have a serious studio facility for mixing, and several programming rooms for the writing/preproduction teams in which they will develop the songs, sometimes in collaboration with the artist, sometimes speculatively. Developing backing tracks in downtime is also a good idea, as visiting artists can browse through the catalogue of these tracks and use them to develop the melody and lyric, and this process can happen both onsite and offsite (for example if the artist is going on tour and plans to write melodies and lyrics between gigs).

This production/publishing stable model has been successfully used by teams like Jimmy Jam and Terry Lewis, Cutfather and Joe, Merlin, Babyface, Dr Dre, Rodney Jerkins, Stock Aitken & Waterman and many others to create a myriad of hit songs for hundreds of artists.

Many successful publishing houses are also using a similar preproduction and creative development model to great effect, for example, Peermusic UK, who signed Newton Faulkner for publishing and set up a production company with him, gave him studio time/facilities and an experienced engineer/producer to assist in the development of commercially recording his songs, and after two years' hard work, successfully licensed the recordings to Sony BMG. The album 'Hand Built By Robots' then went platinum.

What's good about this approach is that the artist has a greater degree of creative control over the recordings and owns the recordings, and the producer/publisher gets to develop a diverse catalogue with many covers by different artists on different labels, each cover presenting an opportunity for success. It frees the producer/publisher from the cash-consuming problems of commercially marketing the recordings (which is what the major labels do best), and helps the major labels, who can then focus less on the A&R process and more on the marketing and distribution of the recordings. It is worth bearing in mind that sod's law plays a regular part in the marketing and distribution of records, i.e. 'If it can go wrong, it will go wrong'. So the idea of working for several years on just one album project for a producer can be very risky, because when you hand over the finished master recordings to a record company virtually anything can happen and usually does. Major labels are notoriously fickle and are also quite territorial, for example, if you sign a new artist's project to a worldwide album deal with a major record company in England, and the release of the first single goes badly wrong, chances are that not only will the label not bother to release a second single, but they will probably shelve the album as well. Worse still, because the 'home territory' release failed, the other subsidiary offices of the major label will not go of out of their way to rescue the project, as they have their own local priorities and tend to only service international hits from their subsidiaries. For a young producer with a developing career there are a variety of options:

- You could spread yourself around, working with a variety of artists and labels.
- You could focus on two or three key tracks (i.e. singles) on people's albums instead of whole albums, freeing yourself up to work with more artists.
- If you are the owner of the recordings (as a production company), consider 'territory by territory' licensing, finding the best large independent label partner in each key territory to work with on the project, rather than a global deal with a major. This model will need more supervision, but circumvents the fickle political and territorial problems of working with majors.
- Only sign global major label deals if they show true commitment (i.e. guaranteed releases) and invest heavily in the project — they will want to recoup their investment.

Diane Warren is probably the most prolific songwriter of our time, and has had hundreds of fantastic hits with countless artists all around the world. She controls her music through her own publishing company, Realsongs (EMI). Diane has dedicated her whole life to songwriting. She is a fine example of determined dedication to her art; one example of this was Cher's single 'If I Could Turn Back Time', from her successful album 'Heart Of Stone'. Diane had written the song for Cher, and knew that it would be perfect for her. However, when Cher first heard the song demo, she didn't like it and was not keen to record it. Diane, at this point, took things to another level and grabbed Cher around the leg, refusing to let go until she relented and agreed to record it! Cher finally agreed and the rest, as they say, is history … it was a worldwide smash hit, reaching No. 1 in Australia, No. 3 in the USA and No. 6 in the UK.

In 1996 I had the pleasure of seeing my friend David (Hawk) Wolinski feature as a panelist at the UK music conference In The City. It was the songwriting panel and he shared the stage with another of my friends, Graham Gouldman (10 CC), and Neil Finn (Crowded House). Hawk's song 'Ain't Nobody' had been covered again, this time by LL Cool J, and had featured on the soundtrack of the movie *Beavis and Butthead Do America*.

The single was No. 1 (again) in the UK charts at the time (Hawk always refers fondly to this song as "the gift that keeps on giving"). Hawk had had a spectacular 12 months, because another song he had written with Chicago drummer Danny Seraphine for the 1979 'Chicago 13' album, entitled 'Street Player', had been sampled by dance artist The Bucketheads and released as 'The Bomb (These Sounds Fall Into My Mind)', and this track had also been a No. 1 dance single and Top 5 chart hit. Move on 14 years to May 2010 and I met up with Hawk again at the BMI radio play awards in Los Angeles, where he was being given the songwriter award for Pitbull's huge Latin cross-over hit 'I Know You Want Me (Calle Ocho)', which reached No. 2 on the Billboard Hot 100 singles chart, and the video (on YouTube) attracted over 110 million views, making it the most viewed 'electronic' music video of all time and also the fourth most viewed music video of all time. Once again, a 'Street Player' sample had been used for Pitbull's single, this time the horn riff, and the track had become a smash hit; one of the most radio played singles of the year. With a wry smile I remembered Hawk, at the 1996 In the City songwriter panel, where he was asked: "What advice would you give to aspiring songwriters?" and Hawk cheekily replied, "Keep sampling my stuff!"

Change in the digital space 12

In June 2010 I attended a music conference, 'The Great Escape', held in Brighton, England, with a focus on the artist and live music. The conference featured discussion panels based around the changes in the industry, reflecting on 2009 and looking at what new entrepreneurial business ideas and developments were taking place. There was an emphasis on live music and the effects on the artist and the manager as well as the industry at large. The event kicked off with PRS for Music providing some critical evaluation of the UK music industry, led by their chief economic advisor, Will Page, and his colleague Chris Carey. PRS for Music has challenged and assisted the industry by providing critical data.

PRS FOR MUSIC — FINANCIAL RESULTS SNAPSHOT OF 2009

PRS for Music (UK), which represents 65,000 songwriters, composers and music publishers in the UK, collecting and distributing royalties on their behalf, announced (2009 released figures) a 2.6% rise in annual revenues to £623 million.

Tougher trading conditions in the UK music market were offset by strong growth internationally and by the licensing of new online services:

> *Significant increases in revenues from British music use abroad were highlighted (up 19.4% to £166.9m) buoyed by both exchange rate gains and increased licensing activity in new and established territories. Online revenues grew 72.7% to £30.4m, reflecting the increased number of legal licensed digital music services available in the UK and across Europe. This growth (£12.8m) outperformed the decline in traditional CD and DVD formats (down £8.7m) for the first time, though the legal online music market is still comparatively small. Public performance revenues increased by 2.4% despite a reduction in licence fees for small businesses to £44 per annum as more businesses take greater advantage of the benefits of music.*

GUARDIAN REPORT BY VICTOR KEEGAN

Victor makes some bold statements: "In fact it is easier to make the case that the music industry, far from imploding, is one of the great success stories of the recession". His opinion is supported by figures that show what kids are not supposed to be buying. "Last year (2008 figures) sales of singles soared to an all-time record of

152.7m units, an astonishing 33% rise in a year when the whole economy (GDP) contracted by 3.3%". As singles have in the physical world been primarily bought by kids, these same youngsters who are also likely still downloading free music from the internet were still prepared to pay up to £3 a pop for a ringtone. Victor hits the nail on the head when he explains why this is probably happening. "Because there is an easy payment system on phones which didn't exist for some time on the web. Now there is an easy payment system (iTunes et al.) on the web they are starting to pay again. If the big music companies had spent their energies dreaming up a payments mechanism for web downloads instead of suing their customers they could have swept all before them. Instead they were like the crew of a sinking boat that blames the sea instead of trying to mend the leak".

Victor goes on to explain: "Even now practically everyone I meet from the music industry protests that it couldn't be expected to combat the technological disruption that was eroding its traditional model. What piffle. Lots of books have been written about disruptive technologies. They can't say they weren't warned. As it turned out, pretty well every system for monetizing music — iTunes, Spotify, We7, Shazam, Nokia's Comes with Music et al. — has come from outside the industry. What a missed opportunity".

Victor goes on to discuss that PRS for Music has demonstrated that there has been a fall in sales of albums — down from 133.6m units to 128.9m last year, not helped by the closure of key UK retail chains Zavvi and Woolworths — but growth is happening elsewhere in sponsorship, live shows and merchandising, where there is something of a boom happening in Britain. Overall, the music industry grew by an amazing 4.7% in recession-ridden 2008, and will probably be resilient when the full 2009 figures come in. A key fact is that last year income from live music overtook that from recorded music for the first time. He says: "Don't think tracks, think music".

Victor goes on to state: "The future lies in capitalizing on the whole musical experience, as the admirable Music 4.5 initiative well knows: it seeks to bring together artists and entrepreneurs to plot the future. If the quality of the five-minute pitches made at the conference by budding businesses is anything to go by, the future is bright. I loved the way Songkick.com is moving beyond Last.fm by linking songs you and your friends like with information about the band's past and present gigs, enabling you to talk about them after the show. MusicGlue offers free downloads in exchange for e-mail addresses which, over time, will produce geographic patterns showing where there is a dense enough cluster of fans to justify a gig. CloseCallMusic encourages people to interact with live music as it happens, while TuneRights is trying to crowd-source the financing of records. Audiofuel, which matches music to your jogging beat, aims to be the new Ministry of Sound. I loved what Decibel is planning — to have a vast database of meta tags so you can find out details of each member of the band: that Jimi Hendrix played as a session man on a Little Richard track, for instance. That is just the sort of value-added that will lure people away from free downloads".[1]

---

[1]Thank you to Victor Keegan, courtesy *Guardian* newspaper. www.guardian.co.uk/technology/blog/2010/mar/12/demise-music-industry-facts

I concur with some of what Victor is saying but on one thing we differ, and that is that most people's understanding of the music industry does not go as far as appreciating the different income streams that an artist (on the one hand) and composer (on the other) receive. A composer can receive little benefit from 'live', whereas an artist benefits greatly, and so the catch-all phrase 'Don't think tracks, think music' doesn't hold up, in the way I think he means.

A SNAPSHOT OF THE IFPI REPORT – MUSIC INDUSTRY 2010

- **Effect of piracy** – Piracy is now causing the collapse of some local industries. In France local new artist signings fell by 59% on their level in 2002. In Spain local artist album sales in the Top 50 fell by 65% between 2004 and 2009. Brazil shows similar data.
- **Legislation** – France, New Zealand, South Korea, Taiwan and the UK have adopted or are proposing new measures requiring internet service provides (ISPs) to tackle mass copyright theft on their networks.
- **ISPs** stand to gain from partnerships and see revenues in excess of 100 million.
- **Global music market** – This fell by 7.2% in 2009, but within this figure growth was shown from 13 countries, hidden by the huge downturn in the USA and Japan.

RECORDED MUSIC SALES 2009

- Global recorded music trade revenues totaled US $17.0 billion in 2009, a decline of 7.2% on 2008.
- Physical sales continue to fall (−12.7%), a slowing decline from −15%.
- A shifting sales base from physical to digital revenues grew by 9.2% in 2009 to US $4.3 billion (more than 10 times the digital market in 2004). Digital channels now account for 25.3% of all music sales.
- Performing rights revenues (digital) also grew strongly by 7.6% to US $0.8 billion. This reflects an unbroken trend of growth since 2003. Revenues from the sector now represent 4.6% of the total recorded music industry.
- Six markets experienced increase digital sales that exceed the amount of physical decline: the UK, India, South Korea, Thailand, Mexico and Australia. This is a landmark point.

DIGITAL MARKET DEVELOPMENTS

Over the last five years we have seen a tenfold increase in the number of digital retail partners and a fivefold increase in the methods by which consumers can consume music.

Rod Wells, Senior Vice President Digital, Universal Music Group International[2]

---

[2]International Federation of the Phonographic Industry (IFPI) Report 2010, p. 6.

Issues being faced in this changing digital climate

- How the customer consumes music.
- Putting the customer first.
- Listening to the customer.
- Policy making that doesn't alienate the customer.

The more I think about it, while the work of UK Music is an undeniable step in the right direction I wonder if the industry also needs a strategic team focused on technological innovation and copyright licensing, to help the industry keep pace with change; a forward-looking strategic innovation team to ensure that our core structures are better prepared rather than taken to breaking point all the time, finding that we are always chasing our tails. The industry has a moral obligation and an inherent contractual obligation to protect the copyright owners' work. The word 'protect' has to mean in the present and in the future, and you cannot do that job properly if you are not prepared for where the future might lead. What's the alternative? Government interference, potentially creating a framework that could exacerbate industry growth and consumer experience.

Some practical issues to consider

- Auditing of digital distribution companies — Who will be the first company to audit Apple? Apple's first mover advantage with iTunes has provided a fierce competitor (or is it an ally?) to the industry. Apple exerts substantial power over the music industry and so how might the industry respond when it is time to audit such technology partners? Didn't Apple threaten the industry some months ago in turning off the download market? This is power and domination and the industry has little influence, I fear.
- Transparent industry accounting formulas with new technological partners or business models (a bit like the food industry and labeling) — Those who create the product have a right to know how their music is being consumed and the formula being used. Why so much secrecy? The lawyers will all say, that's what your contract is for. But the detail, the methods and the formula, is never detailed in a contract; the writer or artist gets a percentage, but of what? The music publishing world is much better than that of record labels at being transparent as to income and accounting.
- Accounting income to writers and publishers from subscription models, streaming and others — Can they work? Labels are cursing the micropayment accounting culture.
- Artists and managers want clarity on all business models, but labels are far too secretive.
- Educating the public and the industry — Strategic local and international initiatives?

I'd like to bring in some research done by one of my students here. Could the answer prove to be this simple?

COULD A BASIC MUSIC INDUSTRY EDUCATION BE A KEY MOTIVATING FACTOR IN DETERRING PEOPLE FROM ILLEGALLY OBTAINING MUSIC?

Georgia completed her dissertation in summer 2010 and discovered some very interesting information. She was trying to determine whether a basic music industry education could deter people from wanting to obtain music illegally. Her findings proved this to be true in the majority of participants. She found that:

- The participants did NOT think music should be free.
- Access is replacing the need to own music.
- A basic music education about the effects of piracy changed attitudes and acted as a deterrent.
- On balance, music students prefer to download their music.
- On balance, non-musical students prefer to purchase CDs.

Georgia Frost

However, because music education did not deter all pupils it suggests that owing to psychological and economical factors behind piracy more than one deterrent needs to be used to fully combat this problem.

Bear in mind that we have always lived with some form of music piracy, just as shops live with a certain amount of shoplifting. It will never be eradicated. But could the music industry be more involved in music education at school? Could it be the determining factor in changing the attitudes of society towards illegally obtaining music on mass? It's definitely worth the industry taking a serious look at this! I realize there are a number of worldwide educational initiatives headed up by the IFPI, but could we in the UK get the key information into the curriculum? Georgia focused on discussing the impact on the artist and the composer (not the industry), and in the mindset of the young person she clearly made it personal (the situation resonated with them). They didn't want to see their favorite artist hurt or disappear.

I asked industry colleagues at the Academy of Contemporary Music to comment on changes within the following sectors of the music and media industry.

GAMING – INTERACTIVE ENTERTAINMENT, PAST AND FUTURE

This industry sector has seen serious growth in the past 10 years. The music industry has been one of the main beneficiaries of this growth. Audio was first used in computer games in 1972 with the launch of the extremely primitive tennis type game *Pong*. This was followed throughout the 1970s and early 1980s with titles such as *Simon Says*, *Asteroids*, *Tetris*, *Pac Man* and *Space Invaders*. Music started making an impact in the mid-1980s when both Michael Jackson and Journey released games that used digitized music.

Recorded/licensed music first came to the fore in 1995 when Sony released the 32 bit Playstation with a 24 channel sound chip that could provide CD-quality stereo sound. The first game to utilize the music capacity ability of the Playstation was Quake in 1996 with a soundtrack written by Trent Reznor from US rock band Nine Inch Nails. What followed was that music became an integral part of every game. With the release of the PS2 in the year

2000, music in games had reached its peak, with soundtracks and scores being exclusively written for titles and music licensing for games such as *Grand Theft Auto*, with its numerous genre radio stations being one of the most innovative ways of getting music into games. In 2004, Sony released the first in its series of *Singstar* titles and the music-based games genre started to gather pace and sales which way outstripped what albums were doing. *Guitar Hero* (Activision) and *Rock Band* (EA) made significant inroads in this area and also gave the record companies a much-needed boost to dwindling catalogue sales.

Fast forward to today and these titles have become household names, with major record companies desperate to have their artists' music on these titles: Take That (*Singstar*), The Beatles (*Rock Band*) and Metallica (*Guitar Hero*). As an income stream and marketing tool the music industry has started to use the games industry as an extremely effective part of the marketing mix in the release of conventional albums. The interactive entertainment industry is still very much in its infancy and as new innovations and developments are announced every year, the music industry has to integrate itself into this industry as a means for survival. Consoles have now become the entertainment hub in the living room with them now having the ability to play your whole music collection, download albums, and exchange music and views via social networks such as Last FM and Facebook. Late 2010 saw the launch of Project Natal (Microsoft) and Playstation Move (Sony), which will change the way we play games forever; a console that can track your every move, mood and music taste. The games industry's issues and challenges for the future are similar to those the music industry has faced for years – the demise of physical product, the preowned market and piracy. The games industry has put strategies in place to combat these problems, and hopefully these will work.

Ewan Grant (2010)

IMPACT OF THE DIGITAL SPACE

The musician and writer Steve Lawson recently defined the changes the music business is facing as transitional. Transitional as opposed to incremental. Historically, the industry saw technological change as an opportunity, for example an opportunity to sell music in a different format and the incremental financial upside realized when consumers moved from vinyl to CDs. However, the digital landscape has completely transformed the marketplace. Incremental change is evolution. Transitional change is revolution.

The digital revolution has wiped most high-street music retailers off the map, has seen a generation grow up with the idea – if not the reality – that music is free, empowered them to create their own content, create their own music, market and promote it, share it and be social with it. It has seen the rise of the track and the decline of the album as consumers choose the song over the bundle and it has changed how these same consumers access media, changed the role of traditional gatekeepers and changed the relationships artists can have with fans.

It is worth noting that YouTube is only five years old, Facebook four years old (counting from when it went completely open) and iTunes started in the UK only six years ago.

Yet each of these three companies has had a major impact on society, media, technology, consumers and the industry in a remarkably short time. Overall it was an impact the industry was ill prepared for. Traditionally the industry, either through vertical integration or just by holding the purse strings, had controlled production, manufacturing and distribution, and heavily influenced access to mainstream media. Over the last 10 years the consumer – and affordable and easy-to-access technology, whether via websites or software – has taken that control away from them.

For artists and curious music fans every week brings new sites, technology and metrics to develop sophisticated marketing, distribution and consumption offers and solutions for minimal expenditure: sites such as Bandcamp that offer artists a complete solution to market and help to distribute their music in a personalized way that traditional retail could never match; Soundcloud, a company that – as it simply states – 'moves music' effectively across the web with an interface that is both effective and intuitive; and MSpot, a competitor to Spotify that offers consumers the opportunity to store their music collections in the Cloud and access them on the move.

The opportunities offered by major labels to artists were always limited to a very few. A label signing was an unrealistic goal for many artists, yet at the beginning of the twenty-first century artists have greater opportunities and freedom to grow and build sustainable careers. They get to keep ownership of their copyrights, to monetize relationships and to take control. Independent labels see this transitional change as positive, since it has leveled, to some extent, a playing field that was previously 'owned' by the majors. There is still an imbalance at mainstream media, although how important are mainstream media when your core fan base or markets ignore it anyway?

If there is one phrase that illustrates the new era the industry has been, and is, facing it is freedom of choice. Freedom of choice to take risks and launch start-ups, freedom of choice to launch own labels and self-release and, most importantly, the freedom of choice not just to see things differently but to be able to implement that personal vision.

What does this mean going forward? Simply no one really knows yet. We increasingly see the future purely through the prism of digital but it is worth remembering that in 2009 80% of all albums sold in the UK were CDs. There is a tendency – or to be more precise a temptation – to focus on the digital, but physical sales are still a major, indeed the major part of the sales equation. A cursory glance at Amazon Market Place or eBay shows a huge market, albeit secondary, for physical music.

The question I find compelling is not whether digital will replace physical, but how will the two co-exist?

In some cases they will be strange bedfellows. The concept of Cloud technology, for example, is both exciting and challenging. If your future infinite library of music is held in a virtual locker that you can access however, whenever and wherever you like without actually 'owning' what you have bought, what happens when you lose the key? There will be issues of privacy, ownership, quality and, as always, price. The Cloud is not new, just new to music. Most of us already pay a monthly subscription via cable or satellite to access television, films and other content without actually 'owning' it. Is it such a leap to imagine the same with music, or do we have a different relationship with music? In the end, the consumer will decide. In the short term it will be predominantly the major labels that decide whether the consumer will get to make that decision.

The future will be digital, if for no other reason than each generation will take for granted what we see as new. Of that there can be little doubt. In fact, the present consumer experience is increasingly digital from conception to consumption. How consumers and fans engage with music is frankly more important than the method by which they do so. Music is more than ever facing a deficit in, or at the very least competition for, consumer attention, and a declining share of their disposable income.

How we define the value of music is a philosophical as much as a commercial question. Is an MP3 file vested with the same cultural significance as a piece of vinyl or a CD? Sometimes it is the simple questions that are the most challenging. The reason most recorded music is sold in the last financial quarter of the year is because it is bought as Christmas gifts. How do you translate the tangibility of a physical gift and perhaps the emotional message the gift represents to a digital cloud or even an MP3 file?

When I talk to students about the music industry I try and impress upon them one fact. We are living and working in one of the most exciting periods in history. A period when more people have more freedom to take control of their careers, their future, their talent, and to actually do something with it, than at any other time.

The question we seemingly ask ourselves every day is: how are we going to sustain an industry on the back of this new-found freedom? Perhaps we aren't. Perhaps the old music industry is dead or at the very least terminally ill. Can an industry that was built on a hierarchical model survive in a world that is now flat? The answer has to be no. The choice of company name by internet radio service Pandora was very apt. The box has been opened and it cannot be shut. That is both the opportunity and the challenge facing the music industry.

Mike McNally (2010)

BUT DOT-COM BOOM? THAT'S *SO* NINETIES! … RIGHT?

"Excuse me sir, can you tell Chantelle? She keeps on tweeting!" Chantelle scowls. This isn't the first time that her inner entrepreneur has caused trouble within the classroom, but unbeknown to anyone, 17-year-old Chantelle, a heavy-metal music lover from Clacton-on-Sea, is one to watch. From her mobile phone, she's managing a dot-com business with huge boom potential, and she doesn't even know it yet. Neither do her classmates. Asking Chantelle to put down her telephone and getting back to our lesson on e-Business, neither do I.

Usually surrounding drivers of change, there have been plenty of such booms (or bubbles) in history. When the world was young, agriculture drove change in the trading environment and created money-bubbles to match. As the nineteenth century hit, industry gained control (think the 1840s' 'Railway mania' boom, think the coal boom, think the metal boom), and then, in the twentieth century, technology took over.

Say hello to the 1990s and a brand new dot-com bubble. If you were serious about business, this meant websites, e-commerce and global trading. Those who got in there first reaped the rewards of advances in media and communications technologies, and whether existence was short-lived (we must recognize that for some, the bubble burst) or guaranteed (just look at Amazon, eBay and Google), mountains of money were made and businesses of gargantuan size were built.

But what about today? Has the final whistle been blown on media and communications technology, or is there a second bubble still to come? With some big acquisitions having taken place recently (think YouTube by Google and Skype by eBay), mergers of communications giants (think Orange and T-Mobile) happening as we speak, and innovation in technology showing itself daily, I think the answer is, 'almost certainly …'. But, before we all rush to the drawing board in an effort to create the next big thing in the dot-com world, let's get back to Chantelle and her classmates, and question whether we need to be thinking big at all.

Forty or so students sit before me. Yes, a few traditionalists are taking notes down using pen and paper, but the majority (all nineties-born kids) use a laptop. Some of the more edgy guys are using their smartphones. There's even one with an iPad.

They talk, they text, they e-mail, they chat, they tweet, blog, comment and the like on a daily basis; turned on and tuned in to a wealth of information streams that are delivered over socially enabled infrastructures, networks and platforms controlled by the big guys: the Googles, the Vodafones, the Facebooks of this world. And those big guys are making huge amounts of money; from airtime contracts, from access subscriptions, from advertising – but still it isn't enough. So now they're going after other big guys. Google is going into travel (their recent acquisition of travel technology company ITA Software is a clear threat to those such as Expedia). Amazon is going into groceries, but shares in the online retailer Ocado remain some of the most highly sought after by short sellers, three months after the company listed on the London Stock Exchange.

This diversification will ultimately end in less competition (history has taught us that the big guys always wipe out the small fries), less choice for the consumer, and a more difficult trading environment for new entrepreneurs and business start-ups. So how can you make it big without starting big? They say that money breeds money, but the likelihood is that Chantelle and her chums don't have billions in the bank to make something work. Not to worry, though. They already have the answer.

It's holiday time for the students, which means it's marking time for me. I open up Chantelle's e-Business assignment and am wowed by a gleaming example of a new dot-com start-up with oodles of potential for making money. Her executive summary reads like a dream. Chantelle has taken something that she loves (metal music), spotted a gap and created an efficient, electronic business with hardly any start-up budget: a socially enabled and aggregated, magazine-style web portal for the metal music community. As well as news, a forum, etc., she has products and services to sell (like concert tickets, travel packs), suppliers to provide them (like WeGotTickets and Lastminute.com through their affiliate programs) and customers to buy them. Her site is available on multiple platforms: on mobile, on touch interfaces, on computer and on the big screen. She knows the market inside out (she is the market) and she has a very clever strategy for making her millions … it's all about creating a scalable model, potential for growth and an alluring investment opportunity for a big boy. Mainstream companies providing services (like travel planning) for a general audience may find the opportunity to be introduced to a brand new market irresistible.

If you fancy creating some dot-com boom potential of your own, then start with these easy tips:

- Find something you love. Although it may seem a little *hippy*, this is paramount in a project's infancy, where many years can pass before financial reward is seen. If you love your area of business, you will never feel like you are working, and the energy for what you are doing will continue to flow.
- Do your research. It is important that you follow a market-driven strategy, creating a business in answer to market conditions and needs.
- Spot a gap and get in there. Choose a *something* (a product, a service, an experience), a *someone* (a market) or a *somehow* (a route to market) combination that you can call your own. It doesn't matter if someone else is doing one or two of these things already, but the overall mix should be unique to your business.
- Don't think big … think small (but sustainable). After all, you'll never beat the big boys on a mass-market product or service without throwing wasteful amounts of money at it. Go niche.
- Brand it. Work on your branding and identity. Define your project's values and personality. Know its benefits and its positioning in the marketplace. A well thought-out brand strategy is key for any would-be successful dot-com start-up. The added importance here is that while concepts cannot be protected, specific implementations of concepts can. The more developed your brand is, the more intellectual property (IP) protection you can gain, increasing the value of your business.
- Bag yourself a good domain. Be creative. Make sure it says what it is. There are so many domain suffixes out there (with new ones being introduced regularly), don't fall into the trap of thinking that .com is your only option.
- Keep your costs low. The e in e-Business shouldn't just be for electronic, but for efficient. If you can match suppliers and customers, making money by taking a cut of the deal, then do it. You might not need to get more involved. Many companies offer affiliate schemes that pay commission when your promotion of their product/service results in a transaction. So take advantage of third party networks, platforms, infrastructures, manufacturing, products, services and human resources instead of having the expense of managing your own.
- Seek out your market. Communicate with them. Maintain their interest and keep them there. Today's media-savvy consumers want to eat, sleep and drink things that interest them 24/7. Take advantage of efficient, targeted advertising services being offered (such as Facebook Ads or Google AdWords). Use relevant social networking tools and platforms to create a seemingly constant flow of information surrounding your business.

- Know your customer inside out. Know what they like, what they don't, and most importantly, what makes them part with cash. Take a hint from the social networks and use profiling (whether it should be explicit or implicit) to gather information that may be useful to you (and others, at a cost!). Use customer relationship management to exploit that knowledge and promote commerce.
- Know your competition even more.
- Multiplatform is no longer good enough. Think *every* platform. People interact with the online domain in many ways: on mobiles, touch-enabled devices (think iPhone and iPad), laptops, desktops and televisions to name a few. Consider some differentiation, optimizing your site for use over these various platforms.
- People need to know what it is they are getting before they can be expected to pay for it. Don't introduce a complete paywall to your dot-com business, and perhaps even consider implementing some kind of Freemium strategy in order to gain a strong customer base during the project's infancy. This has worked wonders for the likes of Spotify. Free product for free information is a proven and valuable way of learning more about an audience.
- Identify your exit strategy. Develop your business so that the big boys invite you out to play some day. Look at Lala (a little-known music streaming service that was recently acquired by Apple for $17 million). They had knowledge of the market, a working business, a small customer base and huge potential. It is suggested that Apple bought Lala for the purposes of modeling a new incarnation of iTunes based on the very en vogue Streaming/Access based model (watch this space), so think about *your* knowledge, *your* market, *your* business, and who it might one day be valuable to.

But, most importantly:

- Do it and enjoy it. The online world is a fun place for a company to exist. Say goodbye to land and sea borders, goodbye to issues of accessibility, and hello to the excitement of the dot-com business.

Oliver Sussat

GERD LEONHARD — FUTURIST, FATALIST, FANTASIST OR GURU?

Gerd is a futurist and he prompts you to think: is it that which makes his input so worthwhile? Get involved: http://www.mediafuturist.com

His comment: "Boy, is it URGENT to sell access not copies. Even a blind person can start to see this now. Licence ACCESS. Share revenues". He quotes Muserati as:

> ... *Rather like the 'Home Taping is Killing Music' campaign mounted by record companies in the 1980s, the arrival of illegal file-sharing coincided with an increase in legitimate sales of recorded music in the three largest markets: America, Japan and Britain. This supported the file-sharers' defence that their activities were no more harmful to music sales than the arrival of free radio airplay in the 1930s ...*

There is a key fundamental difference between the above and internet piracy: tape copying has several limiting factors. It takes time to do and there are fixed paths to distribution: road, sea, plane, one to one. The difference with illegal file-sharing is that in an instant the file is available to everyone all over the world, to millions in fact. Essentially the scale and ease of file-sharing cannot in any way be considered the same as physical forms of tape copying or illegal duplication and sales. They cannot be compared in any way, including the defense that internet piracy promotes sales: it doesn't. In this respect there can never be a successful solution to just 'licensed access' without also having structure and regulation (licensed access is itself structured regulation to online access and so confirms my point). Anyone also assuming that merchandising and live income provide the other piece of the jigsaw puzzle to our missing billions is also deluding themselves. Composers and song-writers make successful artists, not the other way round. It is the composer and the songwriter who are not protected by such increasing revenues compared to the dwindling share of mechanical income and performance income through decreasing releases. Many composers and lyricists are no longer able to support themselves, and we must all share this blame. Why? Because we have allowed excuses in bad business practice to condone theft. If we are to have some opinions let them be of substance, and combine the 24/7 access and freedom with structure and licensing and no longer give internet theft an easy ride. Many different business models have been hailed as 'the answer' and in truth the answer is what we want to make it. There are always people looking to profit from the misfortunes of others. Thank goodness governments and ISPs are finally joining the debate, reluctantly or otherwise. As a civilized society we have to decide what is right and wrong, what we want to protect and defend, and be able to share and gain from this in the future.

If we value creativity the music composer, the lyricist, the arranger, the musician, the artist, the producer, then we start by understanding that they have separate income streams from each other, and they all deserve a future; they create the products that businesses (record labels, publishers, MySpace, mobile telecommunications companies, film, television, games, agents, promoters, Google, Facebook) profit from.

So when the next great songwriter emerges and we find ourselves enriched by their melodies and lyrics, who will walk up to that person, look them in the eye and say:

> *Give me all that you have written, and all that you will write in the future, for you will never be given the chance to support yourself or your family because I want your music now, without paying you a penny; oh, and by the way, I will give your music to anyone I wish and they won't pay you either.*

Gammons (2010)

What despair that day will bring, what a society we will have become. Well we are there; this time is now! Where does our moral compass lie?

Businesses are working hard to supply what consumers want, pushing the boundaries further as soon as technology allows, and yes the music industry must be far more proactive in licensing access, on this I think we all agree. But there must still be boundaries and not excuses. Consumers must be drawn to facilities that are licensed so that the industry in all its shape and form can protect those who create it.

There is a fundamental difference between giving consumers what they want and condoning theft. Some bold moves need to take place. Cinemas run a copyright protection advert before all films; why not grow this idea across all forms of media? Through television adverts, posters, schools, redefine the boundaries, bring the problem and the solution into the open, reward the public for accessing music through legitimate sites, educate children — our future society.

> *Here is my simplified recipe: (1) Offer ALL music in open, 100% compatible formats. (2) Offer back catalogue deals, bundles, subscriptions, and sooner or later, flat rates. (3) Start providing added values that only YOU can provide (such as bonus tracks, video, chats, blog/backstage access, concert downloads). (4) Start treating the users/listeners/fans with the love they deserve, instead of with the hate that your lawyers have in stock for them.*
>
> **From Gerd's blog (April 2007)**

Well in 2010, consumers have all of the above and internet piracy is running into billions. But, you know, as long as we continue to blame the industry and advocate that consumers should have things if they want them, we are condoning theft! Do we teach our children to just take what they want and to hell with the consequences? Well, I certainly do not.

Without acknowledging the responsibility we have to each other, civilization will start to crumble. I'm sorry, but when did theft become a legitimate way of owning something? When did wrong become right?

The industry made some catastrophic misjudgments with the internet and I was also one of the first to bang the drum in this regard. But now all we are doing is giving oxygen to the legitimization of theft, by this constant handing out of excuses; this cannot be the future. We must all agree to this and change our attitudes. The future must be in providing all the things Gerd mentions and more, and the industry is doing this as fast as it can, but this must now go hand in hand with cutting off (using ISPs and legislation) those who seek to rob our creatives and who will and who are stopping writers from composing, stopping musicians from playing. When we lose this ability to value music and culture we have lost our souls, we have lost our direction, we have lost the point. My comments may rub people up the wrong way but others I know will support them. What's your position? Put yourself in the shoes of a composer and see the world from their perspective.

CHAPTER 13

Purchasing a catalogue

With all the reading that I have done I have never found any written guidance on how to value copyrights. This chapter will pull back this veil of mystery and provide a few nuggets of information with which such skills can be developed. It is an area that has many facets to it and so I will try to give a simplistic overview in the first instance.

One of the best business entrepreneurs who made a successful career out of buying and selling music publishing catalogues was Freddie Bienstock. Sadly, he died in 2009 at the age of 86. I was lucky enough to spend some rare time with him when my husband signed to Carlin as a composer in 2001 and we met to discuss various publishing matters. I will share with you some information about Freddie's career.

During an illustrious career, Mr. Bienstock worked with such songwriters and music business executives as Leiber & Stoller, Cliff Richard, Bobby Darin, Ray Davies, U2, John Sebastian, Tim Hardin, Eric Burdon, James Brown, Peter Allen, William Bolcom, Ernesto Lucuona, Norman Dello Joio, Carole Bayer Sager, Kander & Ebb, Koppleman & Rubin, Marvin Hamlish, Stephen Sondheim and Elvis Presley. According to his company biography, Bienstock worked closely with Presley, who often relied on the publisher to choose the songs he recorded.

After emigrating to the USA before the start of World War II, Mr. Bienstock began his career in the stockroom of publisher Chappell and Company, later becoming its chairman. Bienstock, who served on the National Music Publishers Association's board of directors for nearly 20 years, founded Carlin Music in 1966 by acquiring the Belinda Music catalogue and built it to more than 100,000 songs.

NMPA chairman Irwin Robinson called Bienstock, "More than an icon and leader in the music publishing industry", and NMPA president and CEO David Israelite added, "Freddy's passion for music and commitment to artists and songwriters made him a giant in our industry, and his legacy will not soon be forgotten".

Freddy had always owned Carlin Music. "When I bought the original Hill & Range company in England, which was called Belinda, I changed the name as I wanted to call it 'Caroline' after my daughter. There already was a Caroline Music so I took a few letters off the name and called it Carlin". One of Freddy's first signings was Ray Davies, so he struck gold once again. He says: "I was involved in all Elvis' pictures and I was also involved in *The Young Ones* and *Summer Holiday*. I gave Cliff a very good song, 'Traveling Light' by Sid Tepper and Roy Bennett. I'm surprised that he didn't get established in the States until he did that duet with Olivia Newton-John many years later. We also gave some songs to The Animals and Mort and Leslie McFarland wrote a No. 1 hit for Billy J. Kramer, 'Little Children'."

The Art of Music Publishing. DOI: 10.1016/B978-0-240-52235-7.10013-2

Unfortunately, my time with Freddy Bienstock was drawing to a close as it was time for his lunch engagement. Any interview with him was bound to be unsatisfactory as there was so much to ask him and he was such a good talker. Although his time was spent so productively in looking after his copyrights, I had hoped that he would write his autobiography. "A lot of people have asked me to do that but I haven't really considered it. If I look back, I know I have had a marvelous life and a very interesting one and I've met a lot of interesting people, but no book is planned at present. ... New opportunities are always presenting themselves. After I acquired Chappell in 1984, I became aware of U2 and I started to negotiate with their management. We did this for a year and came to a deal which was a three-album deal for $5 million. I told the board that I had made this deal with U2 and one of the bankers came to me afterwards. He thought that I had made the deal with the U2 pilot who had been shot down over Russia and he wasn't sure what I was going to do with him. I had to explain that U2 was also the name of a group. It was a gamble but as we were publishing all the songs, we were getting 50 cents an album, and within eight months, *The Joshua Tree* had sold 14 million albums and we had $7 million back on a guaranteed $5 million. It happened to be a very good deal'[1]

This was all in addition to representing Jobete Music, the publishing arm of Motown Records. Three years later, Freddy left Hill & Range to form The Hudson Bay Music Company with songwriters Leiber and Stoller and rapidly acquired such major assets as the Koppelman and Rubin music firms, which published the songs of Bobby Darin, Tim Hardin and John Sebastian, and later, the label and publishing divisions of two seminal rhythm and blues companies, Starday and King Records. His holdings were further enlarged by the acquisitions of Herald Square Music and Times Square Music, the New York Times music firms that were the pre-eminent Broadway show publishers of their era with scores such as *Fiddler on the Roof*, *Cabaret*, *Company*, *Follies* and *Godspell*, and hits by songwriters Peter Allen, Marvin Hamlisch, Carole Bayer Sager, etc. When his partnership with Leiber and Stoller ended in 1980, Freddy, along with the estates of Oscar Hammerstein II and Richard Rodgers, acquired Edward B. Marks Music Co., the publishers of George M. Cohan, Billie Holiday ('Fine and Mellow', 'God Bless the Child', etc.), Ernesto Lecuona ('Malaguena', 'The Breeze and I') and Jim Steinman (Meat Loaf, etc.), in addition to numerous classic American song standards and works by such significant serious composers as William Bolcom, Norman Dello Joio and Roger Sessions. Soon afterwards, Freddy acquired and became the largest stockholder and Chairman of Chappell and Co., which he then sold to Warner.[2]

This chapter will look at the fundamentals of valuing a catalogue. The buying and selling of catalogues in the music industry occurs fairly frequently. This chapter will aid publishers who would like a better understanding of such matters, but equally, if you have a catalogue to sell then it is worth knowing what you might be

---

[1] http://www.biwa.ne.jp/~presley/elnews-Bienstock.htm
[2] http://www.carlinamerica.com/about/

able to get for it. If the money you pay for a catalogue is too high, you may not recoup on the deal and make a profit.

What type of catalogues come up for purchase? All sorts. From a small boutique catalogue to something more substantial, which may entail the purchaser borrowing money for the purchase. So it's incredibly important to know what you're buying and how quickly the deal will recoup and be able to estimate profits. The one area of the music industry that the City understands is music publishing. You will find that most media banks will be fully cognizant in such matters especially if money is being borrowed against such assets.

EXAMPLE – WARNER MUSIC GROUP

Warner Music Group (WMG) announced recently that it has acquired three music publishing companies. Under the agreement, WMG's Warner/Chappell Music, Inc., one of the world's leading music publishers, will obtain 100% ownership of the following:

- Megasongs Publishing A/S – This Danish company's catalogue includes worldwide hits written by Herbie Crichlow for the Backstreet Boys (including 'Show Me The Meaning Of Being Lonely' and 'Quit Playing Games (With My Heart)' and O-Town, as well as the songs of multi-platinum selling group Ace of Base (including 'All That She Wants', 'The Sign' and 'Beautiful Life'), among others.
- Dizzy Heights Music Publishing Ltd – A UK catalogue comprised primarily of popular songs from the 1980s and 1990s that includes the first four albums by the influential rock band The Waterboys, who were responsible for the 1985 hit song 'The Whole of the Moon'.
- Glissando Music Ltd – A catalogue of UK hits that includes classics by songwriter Chris Andrews ('Long Live Love' and 'Girl Don't Come').[3]

More examples are given in Chapter 1.

[3]http://www.thefreelibrary.com/Warner+Music+Group+Acquires+Three+Music+Publishing+Companies+From

A company wanting to purchase a catalogue may be motivated by different things. Market share, profits, competition from other parties, gaining press coverage, market confidence (which impacts the share price of a floated company) can all play a significant part. But there is a sound process that should be gone through. It might appear sometimes that some of the numbers used to value catalogues are plucked out of the air, double digits sometimes being used. The valuation process is not the same between a music publishing catalogue and that of a catalogue of masters. If you get it wrong you end up with the scenario that Guy Hands faces at EMI, saddled with a debt because he's paid too much. Neither is it the same if you're having to borrow money to purchase a catalogue instead of taking it out of cash flow. The true cost of your return on the deal and therefore your financial risk has to be viewed by doing a discounted cash flow to arrive at the present value of those cash flows. How much is it really worth in today's terms?[4]

Sometimes I would love to be a fly on the wall and know how the parties set about valuations on some catalogues! The motivating factors can be various, as can

[4]http://www.creativeacademics.com/finance/dcf.html#work

the types of companies involved, from majors to small independents. There are therefore also different people to please. Some will undergo public scrutiny and others are private companies. We're going to take a look at unraveling the music publishing side and look at the basic principles involved.

It's important to be very diligent throughout this process so that you can be focused on what is important. Do not be swayed by an extravagant sales pitch from the seller or the seller's lawyer. Let the facts speak for themselves and don't part with any money unless you've got a good lawyer and an accountant to assist.

Start with some basic reasoning. Research the catalogue being sold, the company or owner and the hits it contains.

WHY IS THE CATALOGUE BEING SOLD OR BOUGHT?

- Is it important to ascertain why? Well, yes it is. It affects the confidence you have in the catalogue owner if you know whether they are not hiding issues that will affect the value. Are they selling because of financial hardship? This in itself is not an issue, but if it is as a result of failing to account to writers, and there is a trail of debt associated with the catalogue (and this could well be the case) then yes, this will affect the value of the catalogue.
- Individuals may be wishing to change direction and leave the music industry.
- They may be planning retirement with no succession plans.
- Tragedy strikes and the catalogue owner passes away and his or her heirs wish to sell. Michael Jackson's family may one day wish to sell the remaining share in ATV music.
- A catalogue owner is actively approached by a company wishing to buy its catalogue or business. It has been a trend in recent years through mergers or acquisition, as a way to build value and turnover, delivering economies of scale, general cost savings and penetration of new markets, including new geographical locations.

MANAGING RISK

Through the due diligence process all the key information that could affect this deal must be asked for and checked, and indeed warranted as being true. In addition, any areas of weakness need to be identified and those weaknesses determined by importance as to their impact on the financial viability of the deal. In essence, you should identify the risks and decide how important each risk is, say on a scale of 1–10. The higher the risk the more important it is that those areas are clarified to your satisfaction and to the satisfaction of your lawyer and accountant.

When considering risk, consider also any economic pressures, business climate, financial liquidity/market trends, political, economic, social and technological (PEST) factors, and the trading of the company.

Due diligence

This can be described as being the care that a prudent person might be expected to exercise in the examination and evaluation of risks affecting a business transaction.

Due diligence will remove anyone relying on 'gut instinct' and will also remove any element of ego in wanting to do a deal. It provides justification, structure and reliance upon which the owners of the company can rely.

The due diligence process should look at the deal in hand and know how the deal is likely to impact on the company (the purchaser). Consider the following areas:

- **Financial** (cost implication) and payment structure of any proposed deal – Financial background, royalty trends, local and international; financial commitments of existing contracts (advances/payments due) and registered accounts; identifying the key copyrights that turn over the most money: for how long are they signed to the catalogue? Where are the income streams being generated? This will have an impact on future trends and value. Mechanical income is in severe decline.
- **Content** – What do you get for your money? Key copyrights, size of catalogue, retention period, genres, style, profile, era, territory? What does the owner consider the average net publisher's share (NPS)? Test this against the key turnover copyrights.
- **Legal framework** – The contracts to be relied upon; key writers, other catalogues warranties, liabilities, law suits.
- **History** – Background information on catalogue, hits, owners, company.
- **Resources** – Staff, money, future strategy, business development, competitive position, training, system integration.

BACKGROUND INFORMATION

Consider, if someone is in a hurry to sell something, what does this tell you about them? How business is conducted is very important to both parties.

- If the catalogue is littered with gems, you can be assured that there will be other parties likely to be bidding so you need to shut that door as quickly as possible.
- If the catalogue has flagged up any concerns, such as titles you've noticed have not been registered, it will focus you on asking further questions as to why.
- If a business fails to collect money that is due to it, what does this say to you about the business or the people running it?

Of course none of the above may apply, but be aware of such matters. If you are selling a catalogue then also be aware that a due diligence process will be conducted and it is always best to be honest and open. In doing so you will gain the respect and trust of those you speak to that all you present is indeed true and correct. But clearly the seller wants the highest price and the buyer wants the best deal they can get.

The majors are so big now that some extraordinary errors are being made, with catalogue owners not knowing what they control, and titles going to No. 1 without the publisher realizing they were owned by them. This is never a good thing.

So, once you've found a catalogue you want to purchase (or you have a catalogue to sell), what do you do next?

It is important to establish where the value in the catalogue lies. Let's assume that there is a catalogue of 800 songs. The first question to establish as a guide at this stage is: what is the net publisher's share (NPS)? This is the percentage that will equate to a value determined by the contract splits agreed between the publisher and the writer when the songs were assigned to the publishing company.

Let's assume the average NPS comes back at 40%, i.e. the contract royalty splits were on average 60/40 with the writers.

Out of every £100 earned (see net receipts and at source income in Chapter 6), £60 will get paid to the writers and £40 to the publisher. Essentially, the £40 represents the publisher's net receipt, i.e. after the writer has been paid.

It is true to say that the writer must be paid his or her share of the income irrespective of who owns the catalogue. So the value (the net profit of the catalogue, before business running costs) can only be based on the publisher's share and what value this works out to be. The NPS is the only 'value' (profit) in the catalogue (you cannot sell what is not yours to sell, so you must never include the writer's income stream in the valuation). Further, each writer's income stream can be different, for example the writer could be on 60/40 for performance income, 75/25 for mechanical income and 50/50 for synchronization income.

The average NPS would be:

$$(40 + 25 + 50)/3$$
$$= 115/3$$
$$= 38.3 \text{ NPS}$$

Consider also over time that the different income streams have increased and decreased in importance and therefore can have a greater or lesser influence on value. What are the expectations of mechanical income streams for a catalogue in five years' time or even 10 years' time? If, as suspected, mechanicals will decrease significantly, then basing a valuation on the past average mechanical income will be flawed; the value of the catalogue will be unrealistically high. You have to bear in mind the effect of such market forces, and check market statistics and trends.

QUESTIONS

What other questions do you need to ask? What affects value?

- What are the top-earning copyrights worldwide and in each territory? Who recorded them? What chart positions did they reach? What was their most recent activity?

- What period of retention or term is left on the top-earning copyrights? Retention is associated with exclusive writer contracts and term with single song assignments.
- Are there any recent or outstanding law suits?
- How is the income stream of the top-earning titles broken down? Are they radio hits, download or physical sales, or synchronization or print income streams?

RETENTION PERIOD

Assuming that overall there are still 15 years left on the retention period, this means that for the next 15 years after the end of the term the publisher will still receive the NPS, but after this time the copyrights will revert to the composer or original copyright owner. So if the catalogue were to change hands what would a new publisher pay for the catalogue? It cannot be a value that exceeds what you would expect the catalogue to generate in the next 15 years. If you paid more you would have make a bad deal. So how much might you pay? The answer has result from the due diligence process, market trends and other such data. Using the same example as above, if you paid 10 times the NPS you'd expect to recoup within 10 years or earlier and have five years worth of income as your profit. If you paid a multiple of seven then you'd expect to recoup within seven years or earlier and have a further eight years to make a profit. Everyone knows that publishing income goes up and down but, in general, if the catalogue is being worked, income will show a steady increase. (It is important to watch market economic factors as they have a huge impact.) So the intention of the publisher who is purchasing the catalogue is key. If they are buying a catalogue that has got some great copyrights in it but they feel it has been largely ignored they may end up with a great investment, paying a price based on just the average data to date, but knowing that with a little effort they can elevate awareness of those titles and therefore increase revenue. What you don't want to do is to pay more than the catalogue is likely to generate in the time you own it for. Major publishers will sometimes pay handsomely owing to competition and market share and sometimes use multiples that make your eyes water. If you're an independent you cannot afford to take that financial risk.

REASONS FOR BUYING A CATALOGUE

- Increased market share
- A stronger foothold in different territories
- Expansion of songs in different genres and eras
- Press and PR: keeping the share price up
- To ensure competitors do not benefit
- The titles would be great for synchronization use: perhaps it is felt that their future worth could be much greater than the figures to date show

- The owner is keen to sell and the copyrights can be bought cheaply
- Internal growth and development
- Spreading fixed overheads over increased sales revenue
- Expanding the scope of product lines into new markets.

PEST (political, economic, social and technological) and SWOT (strengths, weaknesses, opportunities and threats) analyses would be extremely useful to apply. A PEST can be used to clarify the external market and forces at that time and should be taken into account. A SWOT could be used to examine the catalogue in question and also the motivation and reasons why you are keen to purchase. A SWOT of your own company will help to identify capabilities and areas of weakness which can then form part of your strategy for improvement and development.

POSTCOMPLETION PLAN

Have a strategy in place to embark upon once the deal is concluded. Sounds simple, doesn't it? However, on a bigger scale on a merger or large acquisition it is the lack of postdeal strategy that often sees a deal falter or fail to produce the profits expected.

SUMMARY

1. Structure a deal that makes sense to you.
2. Identify risk.
3. Carry out a full due diligence process (use your lawyers and accountants and primary and secondary research).
4. Clarify the retention period or term and calculate value based on NPS but with a view of market forces and economic trends.
5. Use a further valuation method (discounted cash flow) to estimate the attractiveness of such a deal.
6. Minimize risk to your business by thorough examination of the issues surrounding the key copyrights.

A successful negotiation balances the needs and wants of both parties.

Business and strategic planning

14

So, you might ask, why on earth is there a chapter on business models? Well, my dear friends, music publishing is a business, as are all other sectors of the music and media industries, and is subject to the same pressures as all other businesses, as well as having its own specific threats.

These wonderful business models allow you to work smarter not harder. They allow you to peer into the future as well as the past, and help you to see the express train before it hits you. Talking of express trains, for my Master's Dissertation in Business Entrepreneurship in 2001, entitled 'Tightening the net – making the net pay', I researched many areas, conducting primary research to ascertain the size of the potential music piracy issues years before any formal research appeared from the industry. So big was the black hole that I estimated that UK organizations would not listen to my findings. But they were proved accurate (probably as much to my surprise as everyone else's) when the British Recorded Music Industry Ltd (BPI) and International Federation of the Phonographic Industry (IFPI) came out with their own commissioned research.

The wonderful thing about some of these models, although they have their limitations is, if used regularly, they help you to see your business landscape in a 3D image as opposed to forwards and backwards. I can't recommend too highly their regular use; before you know it you'll be using them all the time to digest information, working out likely impact, possible routes forward for your business, the likelihood of issues in a given deal structure, and planning accordingly. Know that there is risk in every deal, and be prepared so there are no surprises later.

I am not suggesting for one minute that to use them will prevent business problems, just that using them will better prepare you so that you can make informed decisions.

SWOT – STRENGTHS, WEAKNESS, OPPORTUNITIES AND THREATS

Focus on your strengths, and turn your weaknesses into strengths or minimize their impact. Plan for opportunities and the new business that could follow.

Identify the threats: is the risk/reward ratio in the right proportions? Where are the threats coming from and what can you do to mitigate them? Are they financial, structural, staff based, competitor based? Identifying your threats allows you to plan.

Put simply, if you don't do this spot-check regularly, you may be too blinkered in business to see problems before they arise or miss out on opportunities that have passed you by.

The Art of Music Publishing. DOI: 10.1016/B978-0-240-52235-7.10014-4

PEST – POLITICAL, ECONOMIC, SOCIAL AND TECHNOLOGICAL

A good PEST analysis will look at the industry around you and determine what is going on in this sphere that may affect you or your direction and future ideas.

- **Political** – We've just had a change of government in the UK. What changes in policy making may affect you with regard to its new programs or cuts in budgets? The Gowers Report (brought in by the previous Labour administration), Digital Britain and the role of internet service providers (ISPs), grants and support to the industry, export initiatives, and everything right down to local control through councils and local policy making can make an impact.
- **Economic** – For any business, in the music industry or otherwise, money is in short supply. What is the price of accessing a loan or overdraft? How will a down turn affect the work that I do? Will I need to change my pricing structure due to economic pressures? Keep an eye on world issues: what is happening in Europe and America that may impact on my business now and in time?
- **Social** – Look at trends within and outside the music industry. What are consumers talking about, thinking about, being influenced by? Can you use this information to help you to plan what you do next with your business or product? We are currently seeing a general trend towards and building awareness of 'green' issues. AIM (a UK organization) is commissioning research to look at the viability of labels and managers submitting products digitally to radio, magazines and others, as opposed to having to press CDs, etc. Should the report's findings shed light on such issues, you could be an early adopter and be ahead of the curve. The business school I head is involved in this research.
- **Technological** – This is an ongoing area, and you should consider many aspects and how they may affect you. These may include ISPs, piracy, legislation, Cloud technology, licensing of new rights, impact of digital marketing, blogs, forums and maximizing the strength of the internet.

AIMS AND OBJECTIVES

Let's start by defining what they are. An aim is often something a little vague. I want to be successful before I'm 25. I want to have two games synchronizations and two songs used in television commercials this year. Objectives are definitive steps you take to achieve your aims.

SMART objectives are simple. The objective is the starting point of the marketing plan, once environmental analyses have been conducted. All businesses need to set their own objectives for the products or services they are launching. What does your company, product or service hope to achieve? Setting objectives is important; they keep you focused, with a vision for where you are heading and how you plan to get there.

A simple acronym used to set objectives is SMART, which stands for:

- **Specific** — Objectives should specify what they want to achieve.
- **Measurable** — You should be able to measure whether you are meeting the objectives or not.
- **Achievable** — Are the objectives achievable and attainable?
- **Realistic** — Can you realistically achieve the objectives with the resources you have?
- **Time** — When do you want to achieve the set objectives?

Plans are nothing, planning is everything.

Dwight D. Eisenhower

Planning comprises the working out, the discipline and the execution of a plan. Success is in the detail.

THE MARKETING MIX — THE FOUR Ps OF MARKETING

The four Ps of marketing are product, price, promotion and place (Figure 14.1). A marketing strategy that is focused on getting each of these elements right will help to ensure success.

- **Product** — The product (physical or a service) is what people buy. Market research will help you to adapt your product to what people want, as this will

FIGURE 14.1

Marketing mix: the four Ps of marketing.

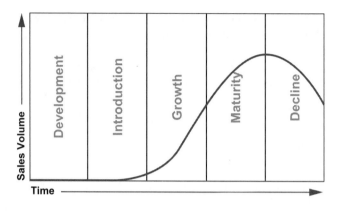

FIGURE 14.2

Product life cycle.

change over time. The time-frame can vary and it's important to monitor both positive and negative elements, and feed the results back into the company.

Every product has a life span. It follows a fairly predictable path: start-up, introduction phase, moving into growth, growth reaches a peak, then the product/service moves into decline and income reduces. This is called the product life cycle (Figure 14.2). When a product moves into the growth phase that's where you want it to stay. To achieve this (look at Apple as a good example) revised models come out, with new features to keep and sustain customer interest but to supply something new that fulfills a need, which exists or is created. The relationship between what you spend on marketing and the return you get for it at any point in the cycle is key. If marketing costs exceed return then you have a big problem. It is important that you ascertain what elements of your marketing plan are driving attention and sales.

- **Price** – Your pricing strategy needs to take into account your costs in securing or providing that product or service, how much profit you need to make for a healthy business, what the competition might be doing and how much the market can stand. Is the product unique in terms of carrying an additional value that could be monetized? How price sensitive is your product or service?
- **Promotion** – How are you going to get your message across effectively to the target audience in the best possible way, being conscious of costs? The key messages need to be defined well, and be clearly understood for that target market.
- **Place** – Ensure that what you have to offer is in the right place at the right time. For instance, if you are selling music products it would be disastrous if on the day the product is released physically or digitally it was not available for any given reason. In music publishing terms, you have supplied a song title to a film and it

is accepted for use, but you have failed to clear such use or failed to deliver the master version — then you have a big problem. All the effort that you had put into your marketing of your music will have failed at the final hurdle and everything done to date would have been a waste of time.

SCENARIO PLANNING

Scenario planning is all about 'what if', a war-gaming strategy used by the military and an effective business tool to broaden and deepen one's analysis of each strategic option. It is the process of creating possible future stories that may affect your small business. It is forward looking and helps to prepare your business for the unthinkable.

Scenario planning came to people's attention in the 1980s when Peter Schwartz of Royal Dutch Shell created one simple scenario: what if oil prices dropped dramatically? Shell thought the unthinkable. Then in 1986, when oil prices were cut in half, the company prospered while other companies were struggling.

Benefits of using scenario planning

- **Early warning** — The majority of winning business strategies result from awareness of an impending change. Knowing when to push forward an idea for your business, change direction, and perhaps pull back from a previously adopted strategy, keeps your business alive and flourishing. In this turbulent business climate and under worldwide economic pressures, it is good to keep reflecting on your planned strategy and different scenarios that may have an impact due to economic influences, political or legislative changes, financial and resource matters. There is always a lot to be thinking about. Take a look at BP: did anyone look at the worst case scenario and plan for a safety back-up that ensured that there was a solution to all major risks? It would appear that scenario planning in such high-risk/high-reward businesses is too blinkered as profits have made them blind to issues of risk.
- **New opportunities** — As your competitors struggle with changing conditions, you will have an edge as you have clearer vision for your business in order to survive and thrive.
- **Risk reduction** — If you assume that your business is safe, wrapped in cotton-wool and impervious to market issues, your business is likely to be in the high-risk category of being susceptible to being fatally wounded. Preparing for possible changes can reduce overexposure of capital to currency markets, and other resources to uncertain risks.

Begin the process with a specific decision or issue around your business: what if US dollar to sterling rates start falling quickly? How might this affect your business if you buy and sell products to that market? This has affected my own business recently; spotting a trend, we implemented new policies and safeguarded the business from these issues.

- **Locate the driving forces** — Discuss what forces will shape the issue. Forces can come from four areas: technology, politics, society and economics. Discuss possible scenarios and likely actions on your part. Can you turn the issue around?
- **Rate the forces** — Some forces will have a greater impact on your issue. A weak American dollar will be impacted more by economic and political forces than by technology.
- **Create scenarios** — Once you have an understanding of the issues and forces, develop several worst and best case scenarios.
- **Discuss implications** — Present your scenarios to your team or discuss them with your accountant or management to brainstorm ideas.
- **Develop indicators** — This is the step where the scenario planning meets reality. Look for key indicators that have to take place for this scenario to become true.

By incorporating scenario planning into your business skills, you have the opportunity to be ready for change. Your decision-making skills will become sharper. More importantly, you will manage the risks of operating a business in any business landscape you find.[1]

SUMMARY

The outline of key business models in this chapter can really assist you in getting a successful business off the ground, and developing a business that has a better chance of growing and surviving the initial start-up period. Most businesses fail within the first two years of trading.

Too many creatives think the music is all they need to rely on. You may have the best song ever, and no one may know about it. You may be marketing your products really well but not seeing the income stream arrive on time or at all. You need to expand and make plans about how you are going to achieve this. No matter what the issue is, this chapter will help you to focus in on the detail as well to explore the bigger picture. Use the models alongside each other and feed the results back into the business and your decision making and strategy.

---

[1]http://sbinformation.about.com/cs/newseconomy/a/scenario_2.htm

Corporate social responsibility in the twenty-first century — making the case for business responsibility in the creative industries

15

Rod Aaron Gammons, Managing Director, Appleworld Distribution

The music business is a cruel and shallow money trench, a long plastic hallway where thieves and pimps run free, and good men die like dogs. There's also a negative side.

Hunter S. Thompson

DEFINITIONS

- **Corporate social responsibility (CSR)** — Broadly speaking, CSR refers to "a concept whereby companies integrate social and environmental concerns in their business operations and in their interaction with their stakeholders on a voluntary basis", as defined by the European Commission.[1]
- **Sustainable business** — This means "development that meets the needs of the present without compromising the ability of future generations to meet their own needs", as defined by the World Commission on Environment and Development.[2] Sustainable business can therefore be seen as an extension of CSR: companies must not only take into account social and environmental considerations, but do so in such a way that is sustainable in the long term.

INTRODUCTION

Ethics aren't just important in business. They are the whole point of business.

Richard Branson[3]

---

[1]http://europa.eu/scadplus/leg/en/lvb/n26034.htm (2002).
[2]WCED. *Our Common Future* (Oxford: Oxford University Press, 1987, p. 43).
[3]*Business Stripped Bare: Adventures of a Global Entrepreneur* (Virgin Books, 2008, p. 10).

The Art of Music Publishing. DOI: 10.1016/B978-0-240-52235-7.10015-6

We are living in extremely interesting and largely uncertain times. We have recently witnessed the near collapse of our worldwide financial system and experienced as a result one of the deepest worldwide recessions of modern times, while at the same we see reports of record profits by the world's biggest oil companies such as BP (who made $6 billion for the first quarter of 2010, up 135% on the same period last year[4]) who then, weeks later, presided over one of the worst environmental disasters in history, spilling between 35,000 and 60,000 barrels of oil a day into the sea for weeks on end.[5] At the same time, businesses, governments and communities across all sectors and industries face increasingly urgent calls from scientists and environmentalists around the world to make the drastic changes required to tackle global warming.[6]

It is not surprising, then, that as a result of the global economic crisis and the increasing awareness of the need to tackle climate change, many people are perhaps now questioning more than ever the incredible largely unchecked power wielded over us by banks, multinational corporations and businesses generally, and questioning their role and fundamental responsibilities in a twenty-first century world.

Yet despite the numerous politicians, environmentalists and academics who express concern over the current position of corporations in society and the long-term sustainability of business in its present form,[7] there is little or no consensus on what (if any) businesses' responsibilities to society entail, how these should be regulated or enforced, and what can be done to move the CSR debate beyond discussion into meaningful action on the part of business. At the same time as the ongoing debate regarding the responsibilities of business is taking place, we are "embarking on a global cultural revolution",[8] which one can suggest is largely being driven by the phenomenal growth of the internet and its latest social networking innovations, such as blogging and Twitter.[9] Indeed, as a direct result of the growth of the internet, global information exchange now takes place on a scale never before seen in history.

The rise of social networking and the internet has resulted in a significant increase in consumer empowerment, and has given individuals a considerably increased ability to hold corporations to account for their actions.[10] It can be argued that rapid increases in global communication through social networking potentially

[4]Source: BP First Quarter Results Press Release, available from www.bp.com

[5]Polson, J. *Bloomberg*, 7 July 2010 [online].

[6]Monbiot, G. *Heat: How to Stop the Planet Burning* (Allen Lane, 2006).

[7]Bakan, J. *The Corporation: The Pathological Pursuit of Profit and Power* (London: Constable, 2004); Zadek, S. *The Civil Corporation: The New Economy of Corporate Citizenship* (London: Earthscan, 2001).

[8]Elkington, J. *Cannibals with Forks: The Triple Bottom Line of 21st Century Business* (Oxford: Capstone, 1997, p. 3).

[9]Sifry, D. *Blog Usage Statistics and Trends: Technorati State of the Blogosphere — Quarter 4, 2006* (2007), http://www.masternewmedia.org/news/2007/04/06/blog_usage_statistics_and_trends.htm

[10]Menasce, D. *Corporate Social Responsibility and Sustainability in the Blogosphere* (Edelman, 2007), http://www.edelman.jp/img/ideas/csrandtheblogospherestudy.pdf

offer a solution to many of the problems that are currently restricting the advancement of the CSR agenda, and also to the re-education of the general consumer and the impact of piracy and illegal downloads on the music and other creative industries. Indeed, on just Twitter alone, a social networking site in which many businesses actively operate, there are 55 million posts every single day from 180 million unique users: that's 640 every second![11] Furthermore, there are approximately 3.5 million posts a year specifically on CSR on blogs worldwide, equivalent to one every 10 seconds.[12] There is therefore clearly a significant role for social networking to play in the advancement of the CSR agenda and a definite need to understand its potential impact.

Aim of this chapter

The aim of this chapter is not to provide a definitive answer to specific questions, but instead to provide food for thought, pose questions for debate, and put forward a case for CSR and sustainable business that will be shown to be as relevant to the creative industries as it is to any other, if not more so.

This chapter will discuss and examine the case for CSR, provide an overview of the arguments for and against in the typical context of environmental and stakeholder concerns (i.e. staff, employees and impact of business activities on local communities), and consider the impact the rise of the internet and new technologies may have on the creative industries and in advancing the CSR agenda.

It will also develop the idea of 'business sustainability' within the specific context of the creative industries and make the argument that those operating within the industry today have a 'social responsibility' to find new business models and pricing structures, and increase efforts to combat piracy, in order to guarantee the sustainability and longevity of the creative industries for future generations of artists, managers, record companies, filmmakers and others operating in this sector.

WHY DOES CSR MATTER TO YOU?

So what relevance does a chapter on CSR have to a book on music publishing? Well, perhaps a lot more than you may at first have thought. Let's consider just some of these issues briefly now.

- **How you run and conduct your own business** — How you manage your responsibilities to your staff, your stakeholders/shareholders, your involvement in local communities, your business' impact on your local environment and global warming. Is your business sustainable in the long term? What is your business doing to manage its environmental footprint?

---

[11]Source: www.website-mointoring.com
[12]Menasce, D. *Corporate Social Responsibility and Sustainability in the Blogosphere* (Edelman, 2007).

- **The business case for CSR** — If you run your own business or would like to in the future, there is demonstrable evidence not only that CSR is morally important, but that "it's just good business" (*The Economist*, 2008). In particular:
  - A European study found that 87% of employees feel greater loyalty to a socially engaged employer.[13]
  - A worldwide study of 25,000 people found that 56% of respondents found their company's CSR profile important in forming opinions of that company.[14]
  - A review of 167 peer-reviewed academic studies over 35 years found a positive link between the CSR record and financial performance.[15]

How to engage and operate responsibly within the music industry

How do you operate and conduct your business? As someone working in the creative industries, what responsibilities do you have to help ensure the long-term sustainability and survival of the industry?

- **Behavior/profile of company you spend years of your life working for** — Do you care if the company you are working for is one of the world's largest polluters? Does your company's record of investment and proactive approach in local communities matter to you? What is the responsibility of the company you work for to you and your co-workers, and their families' well-being?
- **Power of the ethical consumer (*you*)** — Consumers such as you can have an enormous power over business through where you choose to spend your hard-earned money. If you don't agree with the ethical practices of a specific organization, are you effectively endorsing their activities by deciding to purchase their products? We all ultimately determine and influence whether these companies succeed or fail.
- **Environmental impact/global warming** — To tackle global warming, everyone must play an active role. What changes, however small, can you make to your impact on the environment? Does it matter?
- **Human rights/equality** — Is cheap international labor, which is likely to have been used to make many of the clothes you are wearing, ethical? Is it right that some companies who operate globally provide completely different human rights and working practices in their developing country operations than they do in their offices based in their home territories, where such practices would not be legal?
- **Personal responsibility** — Where do personal responsibility and CSR begin and end?

As you can see, there are many ways in which CSR and business sustainability are relevant to you and your business, or perhaps in who you chose to work for. Let us

---

[13]Fleishman Hillard (1999).
[14]Choquette and Turnbull (2000).
[15]*Economist* (2008).

now consider more specifically the idea of sustainable business within the creative industries.

TECHNOLOGY AND THE SUSTAINABILITY OF THE CREATIVE INDUSTRIES

While the idea of business sustainability and CSR has typically referred exclusively to environmental and social/stakeholder concerns (i.e. staff and local communities), which are of great importance and relevance, there is perhaps a much more pertinent question for the creative industries: how sustainable is your current business model; how relevant is it to the reality of the modern world and the widely accessible access to 'free', file-sharing, music, film and television content? Furthermore, what is your responsibility as someone involved within that industry to influence, suggest and adopt new models that guarantee the future sustainability for the industry?

We are living in the midst of a technological revolution and with rapid improvements to worldwide broadband speeds and networks, file-sharing and widespread use of iPods, iPads and new media devices, those working in the creative industries face unprecedented challenges to maintain and enforce copyright and intellectual property (IP) rights among the 1.8 billion (and rapidly growing) people now with daily access to the internet.[16]

As technology advances and more people globally gain access to the internet, the greater the threat of IP piracy and therefore income streams and sustainability of the creative industries. If you're a music publisher and/or rights holder, or perhaps an artist or songwriter, do you care if your music is downloaded illegally? Does it matter if your copyright is not upheld and respected? Does this potentially threaten your career longevity or future earnings stream?

The effect of music piracy is clear to be seen: music industry revenues globally reported in the International Federation of the Phonographic Industry (IFPI) Report 2010 declined 30% between 2004 and 2009, and by 7.2% last year alone.[17] As a result, some local music industries are now seeing a rapid collapse — in France new artist signings have fallen by 59% since 2002,[18] and in Spain local artist album sales in the Top 50 fell by 65% between 2004 and 2009.[19] While music consumers' buying habits having changed with a dramatic shift from physical to digital sales, a 940% increase between 2004 and 2009,[20] the music and creative industries generally have yet to adopt a business model and pricing structure that is both sustainable in the long term and offers a genuine solution to the huge and increasing threat of digital piracy.

---

[16]Internet World Stats, www.internetworldstats.com
[17]IFPI Report 2010, www.ifpi.org
[18]Ibid.
[19]Ibid.
[20]Ibid.

Other creative industries such as film and television (and associated music synchronization rights) are now also increasingly coming under threat from the huge problem of piracy and illegal downloading, with the advances in technologies and broadband speeds making it possible to download entire films within minutes and between file-sharers across the globe.

Very important moves are being made to attempt to combat mass scale piracy, for example, in the UK, France, New Zealand, South Korea and Taiwan, the pressure is being put on internet service providers (ISPs) to combat mass copyright theft on their networks by incentivizing them with revenue streams of up to £100 million for the successful protection of copyrights.[21] Yet, the creative industries still need to find new business models, pricing structures and revenue streams that are relevant to the twenty-first century consumer, to effectively stem the flow of piracy and protect the long-term sustainability of the music industry, as well as an industry led re-education of the music/film buying public as to the real effects of illegally downloading copyrighted material.

ENVIRONMENTAL IMPACT

Businesses in all industries now face a call to action to engage proactively in managing and mitigating their environmental impact at a time where scientists and governments globally are asserting the vast importance of tackling climate change.[22]

Specifically within the music industry, efforts are underway following an agreement reached by record companies in 2009 to reduce carbon dioxide (CO_2) emissions from CD packaging by 10% after a recommendation by Julie's Bicycle, an organization seeking to reduce CO_2 emissions in the music industry.[23]

Indeed, CD packaging accounts for one-third of the recording and publishing sectors' emissions and every single CD jewel case contains 1.2 kg of CO_2 emissions.[24] It is estimated that switching to a pure card option would reduce the music industry's greenhouse gas emissions by 95%,[25] yet it is far from being widely adopted.

It is clear to see then that the social responsibilities of businesses operating within the creative industries are as relevant as they are in any other, both with regard to their environmental responsibilities and specifically with regard to the concept of business sustainability from a perspective of combating music piracy and the need

---

[21]Ibid.

[22]Burke, T., Elkington, J. *The Green Capitalists: How industry Can Make Money and Protect the Environment* (London: Victor Gollancz, 1987); Monbiot, G. *Heat: How to Stop the Planet Burning* (Allen Lane, 2006).

[23]ARUP, www.arup.com/News/2009-03%20March/10_Mar_2009_Music_industry_adopts_Arup_carbon_reduction_report.aspx

[24]*Julie's Bicycle, Reducing the Carbon Emissions of CD Packaging*, Executive Summary February 2009 [Online].

[25]Ibid.

and responsibilities of record companies, publishers and artists to find new business models that enable the future survival of the industry at large.

Having established the importance of CSR to the creative industries, we shall now discuss in further detail the more traditional arguments for and against CSR, and the current state of the CSR debate among experts and academics in the field.

INVESTIGATING THE CURRENT STATE OF THE CSR DEBATE

"Corporations are not society's custodians, they are commercial entities that act in the pursuit of profit, not on the basis of ethical considerations ... they are morally ambivalent".[26]

CSR has rapidly moved up the political agenda in recent years. This may be due in part to the realization that the case for responsible and sustainable business has never been more urgent.[27] Indeed, many go as far as to suggest that one of the greatest challenges of the twenty-first century will be the struggle to curtail excessive corporate power. However, despite the numerous number of academics, environmentalists and politicians who believe that there is a fundamental problem with the currently unsustainable role of business in society,[28] there are still many who believe the CSR debate is all hype, or those that still subscribe to the view that "there is one and only one social responsibility of business ... to increase profits".[29] There is also significant doubt about the level of sincerity about those companies that do claim to be socially responsible, and as Hill notes, companies' CSR statements today are practically "interchangeable except for the company name",[30] reaffirming the view that for many corporations CSR is no more than a good public relations exercise.[31]

CSR advocates

Those who subscribe to the school of thought advocating the need for CSR suggest that corporations today are becoming far too powerful,[32] that globalization is eroding local cultures and customs[33] and that business is exploiting workers in

---

[26]Hertz, N. Better to shop than to vote? *Journal of Business Ethics: A European Review* (2001, Vol 10, No. 3, p. 193).

[27]Elkington, J. *Cannibals with Forks: The Triple Bottom Line of 21st Century Business* (Oxford: Capstone, 1997).

[28]Bakan, J. *The Corporation: The Pathological Pursuit of Profit and Power* (London: Constable, 2004).

[29]Friedman, M. *Capitalism and Freedom* (Chicago: University of Chicago Press, 1962, p. 133).

[30]Ibid, p. 185.

[31]Christian Aid, *Behind the Mask: The Real Face of Corporate Social Responsibility* (21 January 2004), http://www.christian-aid.org.uk/indepth/0401csr/csr_behindthemask.pdf

[32]Zadek, S. *The Civil Corporation: The New Economy of Corporate Citizenship* (London: Earthscan, 2001).

[33]Chryssides, G., Kaler, J. *An Introduction to Business Ethics* (London: International Thompson Publishing, 1993).

developing economies. They also suggest that the gap between the rich and the poor is continually increasing,[34] that business is destroying the environment and in the process placing large costs on society. They argue that a delicate balance needs to be struck between the costs to the environment and society of a company producing a certain product, and the benefit to society of having such a product.[35] As an example, in the USA, the annual cost of obesity is now estimated to be more than double the entire revenue of the fast food industry.[36] Given that on average in the USA, one in four people visit a fast food restaurant every day, it is logical to suggest that the fast food industry plays a significant role in causing the obesity problem.[37] Therefore, in this instance, many CSR advocates would suggest there is clearly a large imbalance between the benefit and cost to society of providing cheap fast food to the nation on mass, when society is incurring the costs of the obesity epidemic.

There is a significant body of evidence to support the CSR advocates' school of thought. For example, of the 100 largest 'economies' in the world currently, 51 of them are corporations, a remarkable statistic.[38] Indeed, the annual turnover of General Motors is bigger than the entire GDP of Denmark.[39] It is no surprise then that many suggest that corporations today have become too powerful; indeed, it seems that "corporations now govern society, perhaps more than governments themselves do".[40] If one agrees with this viewpoint, then this is an extremely worrying development as it seems logical to suggest that "corporations ... can never adequately assume the functions of the state".[41] Indeed, the actions of several large multinational corporations would serve to confirm the view that they cannot be trusted as 'custodians of our society'. For example, in one of its many sweatshops in the developing world, official 'Nike' documents and figures confirmed that workers were given just 6.6 minutes and paid just $0.08 to make a T-shirt that would be sold in the USA for $22.99.[42] In this instance, where Nike has largely been left to its own devices to set voluntary ethical standards in its overseas operations, many would suggest it has abused its position of power. This claim is based on Nike allowing a much lower standard of working conditions and pay than

---

[34]Zadek, S. *The Civil Corporation: The New Economy of Corporate Citizenship* (London: Earthscan, 2001).

[35]Sorell, T., Hendry, J. *Business Ethics* (Oxford: Butterworth-Heinemann, 1994).

[36]Schlosser, E. *Fast Food Nation: What the All-American Meal is Doing To The World* (London: Penguin, 2002).

[37]Ibid.

[38]Zadek, S. *The Civil Corporation: The New Economy of Corporate Citizenship* (London: Earthscan, 2001).

[39]Ibid.

[40]Bakan, J. *The Corporation: The Pathological Pursuit of Profit and Power* (London: Constable, 2004, p. 25).

[41]Hertz, N. Better to shop than to vote? *Journal of Business Ethics: A European Review* (2001, Vol. 10, No. 3, p. 192).

[42]Bakan, J. *The Corporation: The Pathological Pursuit of Profit and Power* (London: Constable, 2004).

it would have been legally required to meet in its home market (the USA), where the goods are being sold. This is certainly not an isolated example; there are numerous instances where, as Gibb and Schwartz subtly put it, "good companies do bad things".[43]

CSR dissenters

However, despite the compelling evidence that those 'CSR advocates' put forward, there are many who subscribe to a school of thought that could not be more diametrically opposed. Not only do they completely reject the idea for CSR but they suggest that those businessmen who advocate CSR are 'mindless puppets', preaching pure socialism.[44] Milton Friedman, one of the most influential economists of the late twentieth century, can be seen to be largely responsible for this school of thought, which shall be referred to in this chapter as the 'CSR dissenters'. Friedman famously asserted that "there is one and only one social responsibility of business ... to increase profits",[45] and argued that CSR is a fundamentally subversive doctrine", the widespread application of which "would destroy a free society".[46] Friedman's work has been extremely influential among governments, business leaders, economists and academics – perhaps most notably Margaret Thatcher and Ronald Regan – and it could be argued that it has played a significant role in shaping the nature of modern business. Perhaps it is no coincidence then that some argue that the resulting 'modern corporation' that dominates society today can accurately be described as a "pathological institution".[47]

Friedman's view can in part perhaps be seen as an extension of Adam Smith's much renowned and highly influential work, *The Wealth of Nations* (1776), in which he stated, "I have never known much good done by those who affected to trade for the public good".[48] Smith essentially suggests that those who claim to be acting in the best interests of society, perhaps for example by promoting the adoption of a CSR agenda, would have a much greater benefit to society by pursuing their own personal interest. Friedman and Smith's viewpoints are therefore in direct contradiction with the views of the CSR advocates who suggest not only that the adoption and enforcement of

---

[43]Christian Aid, *Behind the Mask: The Real Face of Corporate Social Responsibility* (21 January 2004), http://www.christian-aid.org.uk/indepth/0401csr/csr_behindthemask.pdf

[44]Friedman, M. The social responsibility of business is to increase its profits. *New York Times Magazine* (13 September 1970), http://www.hec.unil.ch/faculty/afidalgo/MF.pdf

[45]Friedman, M. *Capitalism and Freedom* (Chicago: University of Chicago Press, 1962, p. 133); ibid, p. 120.

[46]Ibid, p. 120

[47]Bakan, J. *The Corporation: The Pathological pursuit of Profit and Power* (London: Constable, 2004, p. 2).

[48]Cited in Rogers, *An Inquiry into the Nature and Causes of The Wealth of Nations by Adam Smith* (1776), Vol. 2, 2nd edn (Oxford: Clarendon, 1880, p. 28).

a CSR agenda is important for modern business, but that it is essential for society, the environment and the sustainability of business in the long term.[49]

So does the viewpoint of the CSR dissenters hold any weight today in a society which some believe "faces extinction"[50] if companies maintain their sole focus on the single 'bottom line' of profit maximization, as Freidman advocates? Well, actually there does appear to be some common ground between the opposing schools of thought. One could suggest that while not agreeing that the only social responsibility of business is achieving a profit, some writers in the field of CSR would perhaps agree to an extent with Friedman that in reality, the legal position of corporations in society today dictates that their sole focus will inevitably be on profit. This is certainly not to say that nothing can or should be done to change this, but nevertheless the position today entails that a company's prime and sole focus will inevitably be on profit maximization. Indeed, it can be strongly argued that the law in its current form across most of the developed world, certainly within the USA and the UK, provides a legal requirement for a company to pursue, "relentlessly and without exception, its own self-interest, regardless of the often harmful consequences it might cause to others"[51] (p. 2). In the UK, the law as detailed in the Companies Act of 1985 states that a company director's legal responsibility is to maximize the wealth of their shareholders and to place the interests of the company first before all other considerations.[52] One may therefore be drawn to conclude that CSR is "illegal — at least when it is genuine".[53] It can be argued that Freidman's view that the only social responsibility of business being to make a profit is technically correct in respect of the current legally defined purpose of business in the USA, the UK and much of the developed world.

However, the obvious key point of difference between the CSR advocates and the CSR dissenters is that the advocates, unlike the dissenters, believe that it is imperative that businesses must change their attitudes towards CSR, and that governments and society should reconsider the legal grounding of the stereotypical modern business in order to force them to formally recognize their responsibilities to society and the environment. They believe this to be necessary to preserve the long-term sustainability of the environment and society at large.[54] This is certainly not to say that we must abandon capitalism and the pursuit of profit, but that we must mold a new form of 'green' or 'sustainable capitalism', abandoning the traditional narrow

---

[49]Burke, T., Elkington, J. *The Green Capitalists: How industry Can Make Money and Protect the Environment* (London: Victor Gollancz, 1987); Elkington, J. *Cannibals with Forks: The Triple Bottom Line of 21st Century Business* (Oxford: Capstone, 1997).

[50]Elkington, J., ibid, Foreword (p. ii).

[51]Bakan, J. *The Corporation: The Pathological Pursuit of Profit and Power* (London: Constable, 2004, p. 2).

[52]See also UK case law: *Multinational Gas* [1983] Ch 258; *Parke v Daily News* [1962] Ch 927.

[53]Bakan, J. *The Corporation: The Pathological Pursuit of Profit and Power* (London: Constable, 2004, p. 37).

[54]Elkington, J. *Cannibals with Forks: The Triple Bottom Line of 21st Century Business* (Oxford: Capstone, 1997).

focus on the single bottom line of profit maximization in favor of a 'triple bottom line' approach of profitability, environmental quality and social justice.[55] Indeed, capitalism already appears to exist in varying forms around the world, for example in Japan where the prime focus of business is to benefit society, and in which profit-ability is seen as a necessary condition.[56]

On the balance of the evidence presented by numerous viewpoints in the CSR literature, one is drawn to conclude that there is a fundamental and urgent need for business and the creative industries to embrace CSR, and to take into account the long-term impacts of business on the environment and society at large may tech-nically be correct with respect to the underlying legal responsibility for business being to produce a profit, but this is not to say that business should therefore ignore any responsibility it may owe to society at large. Furthermore, it is clear to see that politicians and governments are 'switching on' to the CSR and sustainable business debate, and are beginning to create a legal framework in which businesses must act in a way that not only benefits their shareholders and their single bottom line of profit maximization, but does not harm society and the environment in the process. These changes can be seen to be directly in line with those who argue for renewed business focus on a triple bottom line, and with those who assert that we need to remake the modern corporation, certainly in terms of its legal grounding.[57] Indeed, in the UK the amended Companies Act 2006 now legally requires company directors to also take into account "the likely consequences of any decision in the long term"[58] as well as "the impact of the company's operations on the community and the envir-onment".[59] This is a significant step, but there is still a long way to go towards achieving corporate accountability.

CONCLUSION

... there are no magic bullets that will create civil corporations.[60]

One of the greatest challenges we, as a society, face in the twenty-first century is readdressing the long-term responsibilities of business and its impact on the envir-onment and society at large. Indeed, one is drawn to conclude that the argument for the need to embrace CSR is overwhelming. There are, however, several problems restricting the advancement and enforcement of the CSR agenda. To make real progress, it seems necessary to move away from the voluntary adoption of CSR to

[55]Ibid.
[56]Hampden-Turner, C., Trompenaars, F. *The Seven Cultures of Capitalism* (New York: Doubleday, 1993).
[57]Bakan, J. *The Corporation: The Pathological Pursuit of Profit and Power* (London: Constable, 2004).
[58]Companies Act 2006, s.171(1)(a).
[59]Ibid, s.171(1)(d).
[60]Zadek, S. *The Civil Corporation: The New Economy of Corporate Citizenship* (London: Earthscan, 2001, p. 1).

a system whereby business has a formal requirement to act in a socially responsible way, and is therefore directly accountable for its actions.[61] To achieve this, pressure must be directly applied on governments around the world to tighten regulations and legislation in order to give power to the enforcement of CSR.

At the same time, consumers need to apply pressure directly on business and, in doing so, hold them to account for their actions and force them to recognize their responsibilities to society.

It seems most logical to conclude that it is the consumer that has the most important role to play in applying consistent and united pressure on governments to tighten regulations, and directly on businesses to increase their accountability and transparency. The growth of the internet, coupled with the rise of the blog and other forms of social networking, has empowered consumers and individuals around the globe in a way that has never before existed. Indeed, the blog represents nothing less than a cultural revolution and its potential for advancing the CSR debate is enormous. Today's consumers have access to more information than ever before from all corners of the globe, and are able to exchange this information and new ideas freely, at the click of a button. New channels of communication that never before existed have now opened up directly between the consumer and businesses, and between governments and individuals, providing consumers with a voice that has already demonstrated significant power to hold business and government to account on several occasions.[62] There are also many potential benefits for companies in openly discussing their social responsibilities directly with their consumers through blogging, when it is seen by consumers as a genuine commitment to a two-way dialogue.[63] By all predictions, the growth and influence of the blog is set to continue its rapid expansion in the immediate future.[64] The rise of the internet and blogging could therefore prove to be a revolutionary phase in the advancement of the CSR agenda, and all efforts should be made to embrace this new technology by consumers, businesses, non-governmental organizations and governments around the globe. It may also provide a unique medium by which record companies and the industry as a whole can connect with the record-buying public, to educate them on the effects of piracy and illegal downloading, and also to seek a two-way dialogue and advice on finding ways to move forward to protect copyright with tacit support from the offenders and consumers they seek to placate.

It is perhaps somewhat ironic that the future success of the music industry is likely to be largely dependent on technology — finding new methods of copyright

---

[61]http://www.christian-aid.org.uk/indepth/0401csr/csr_behindthemask.pdf; Bakan, J. *The Corporation: The Pathological Pursuit of Profit and Power* (London: Constable, 2004).

[62]Kirkpatrick, D. et al. Why there's no escaping the blog. *Fortune* (24 January 2005, Vol. 151, No. 1, pp. 64–69); Sudhaman, A. Double-edged blog power puts marketers on guard. *Media Asia* (1 July 2005, p. 19).

[63]Kirkpatrick et al., ibid.; Menasce, D. *Corporate Social Responsibility and Sustainability in the Blogosphere* (Edelman, 2007), www.edelman.jp/img/ideas/csrandtheblogospherestudy.pdf

[64]Sifry, D. *Blog Usage Statistics and Trends: Technorati State of the Blogosphere* (2007) http://www.masternewmedia.org/news/2007/04/06/blog_usage_statistics_and_trends.htm; Menasce, D., ibid.

protection, enforcement, deterrent and detection of copyright abuse — when it was precisely a revolution in technology, specifically in new digital music formats and MP3 music players, that was directly responsible for the huge decline in music sales globally.

Those involved in the creative industries have a responsibility to propose, seek out and adopt new business models and pricing structures that are relevant to the twenty-first century consumer in an effort to ensure their long-term business sustainability and stem the flow of revenues lost to piracy. At the same time, the creative industries must continue efforts to lobby ISPs and governments worldwide to act proactively to prevent piracy and flagrant disregard of the copyrights on which their survival depends, and in doing so protect the future generations of musicians, managers, publishers and record company executives who otherwise would not have a career in the industry.

Malcolm Gladwell, in his inspirational work *The Tipping Point*,[65] suggests that throughout everyday life and society, everything from the adoption of specific ideas, philosophies or beliefs to the success of one business or product over another reaches a tipping point when the levels at which momentum for change becomes unstoppable, and that rather than necessarily being a gradual process over time, it can in fact be a rapid process, influenced and carried out by a small but crucial number of influential people who play very specific roles in the process.

It seems likely that in an effort to lay the groundwork for long-term business sustainability and commercial survival in the creative industries, we may well rapidly reach a tipping point in the creative industries with respect to implementing twenty-first century business models and efforts to combat piracy sooner rather than later, as those in the industry are forced to move to protect their interests and in doing so those of others within the industry.

If Malcolm Gladwell is right in his assertion that rapid change can be implemented by a small number of influential individuals, then potentially those reading this book may have a key role to play — what is yours going to be?

---

[65]Gladwell, M. *The Tipping Point: How Little Things Can Make a Big Difference* (Abacus, 2001).

Conclusion

I hope you've enjoyed the book and have found it stimulating and inspiring. The music industry can feel like a vast ocean and a little overwhelming sometimes, but it's actually like a small village and you'll come to realize who it is you need to be working with quite quickly. I feel very privileged to have helped so many young people get started in the Music Industry and to still be involved in an industry that I love. I have no doubt that I will meet many of you at some point somewhere in the world, and I look forward to hearing about what you are doing. This book is something that you can come back to time and time again as you move through your career. There are a great number of legal books that provide a more in-depth legal focus that would complement this book and make a good addition to your reading. Learning is a lifetime agenda, it's never too late to start and you're never too old to start something new. You can find me on LinkedIn and Facebook and I have a website linked to the book coming online: www.theartofmusicpublishing.com

Helen Gammons is available for conference lecture bookings.

Contact: info@syncinthecity.com
Website addresses: www.theartofmusicpublishing.com
www.rotolight.com
www.planetvideosystems.co.uk
www.acm.ac.uk

The website www.theartofmusicpublishing.com is hosted and maintained by the author.

Biographies

ROD GAMMONS (WRITER/PRODUCER/CEO)

Rod Gammons began his musical career as a busy session musician, which led to his appointment as head of percussion studios for the Isle of Wight. Rod's early work as a songwriter and producer followed with labels such as Record Shack, Savage, Dakota, Warner Brothers, Carlin Music and Chappell Music. Rod Moved to London and opened the futurist Berwick Street Studios, working with acts like Ultravox, Jaki Graham, Junior Giscombe, Oleta Adams, PM Dawn, Zoe, Blue Pearl and Imagination. Rod was then hired as UK Head of A&R for leading Japanese label Avex, and he established the Soho Square studios and the Avex America and Avex Asia labels, in the meantime producing 10 CC, Cheryl Lynn and All Saints. Rod bought Alan Parson's north London studio complex and reopened it as Planet Audio Studios, recording acts like The Lighthouse Family, Beautiful South, Mark Morrison and Wyclef Jean. Since then Rod has established himself as one of the leading producers in the UK. Recent productions include the chart-topping smash hit singles from Liberty X, 'A Night To Remember', chosen as the BBC's official Children In Need single, and 'Got To Have Your Love', taken from the multiplatinum album 'Thinking It Over'. Rod also produced and wrote the B-side 'Everything' for this chart-topping single (V2 Records).

Rod has produced, written and engineered many household names including Wyclef Jean (Fugees), Sonique (Universal), Mark Morrison (Death Row) and Gabrielle (Parlophone). Wyclef Jean of the Fugees recently worked with Rod on a music cue with artists Melky & Sedeck for a new Columbia movie, recording and mixing this soundtrack with Rod at Planet Audio Studios.

Rod co-wrote and produced the killer single 'Sista Sista' for Beverley Knight's 'Prodigal Sista' platinum album (Parlophone) and Roachford's single 'From Now On' (Columbia). He also produced Mark Morrison's recent 'Album for Death Row', including the singles 'Playa Hata', 'Innocent Man' and 'Best Friend', which featured the international stars Gabrielle and Connor Reeves. Further credits include various albums and singles with legendary artistes like All Saints, Cheryl Lynn, Oleta Adams, Ultravox, Imagination, Jaki Graham, Worlds Apart, Was Not Was, 10 CC, Bronski Beat, David Sneddon, Michael Ball, Junior Giscombe and David Hasselhoff, to name just a few.

Regular co-writers and co-conspirators include Hawk Wolinski (Rufus), Graham Goldman (10 CC), Michael Jay (J-Lo) and Jon Lind (EWF, Madonna, Vanessa Williams).

Rod has his own production and mixing studio in north London, which is extensively equipped with the latest HD digital technology. Planet Audio Studios has the largest pro tools rig in Europe, together with a flying fader Yamaha DM2000 96-channel digital production console, TC System 6000 Reverb, Dynaudio Air 5.1 Monitoring and a superb Concert Grotrian Steinweg concert grand piano.

Rod has also been a featured composer, creating award-winning scores for a number of Hollywood motion pictures, including the sci-fi thriller *Within The Rock* (Alliance). *The Never Ending Story Part 3* (Warner) contains Rod's song and production 'Fire' (performed by German rock artists ShyBoy).

Hi television and commercials work includes a Compaq TV commercial, a highly successful campaign, and the Australian TV soap hit *Paradise Beach*, which contains seven of Rods songs, which he also produced. Further TV credits include a long-running Seagram's whisky commercial with 'My Discarded Men' written and produced by Rod, which also features Eartha Kitt.

Rod is founder and CEO of Planet Audio Group, based at Pinewood Studios, a specialist Broadcast HD technology supplier (Planet Video Systems, Planet Audio Systems, Appleworld Distribution), and is the inventor and CEO of the Rotolight, an innovative HD light for HD-DSLR and cinematographers. Rod is also a committee member of the British Society of Cinematographers and is helping to organize BSC-EXPO 2011 at Pinewood Studios

ALEXIS GROWER (HEAD OF THE MEDIA DEPARTMENT, MAGRATH LLP, LONDON, UK)

Alexis Grower heads up the media department at Magrath LLP. He has been heavily involved in music matters from the 1980s through to the current time. He is known to work for 'talent' as opposed to the major companies and has an impressive roster of clients stretching back to the dawn of rock 'n' roll, including Roger Daltrey of The Who, Robert Plant of Led Zeppelin, and Motorhead, through to today's current hit makers such as Tinchy Stryder, Calvin Harris, Hurts, Diagram of the Heart and Magnetic Man.

ANTHONY HALL

Anthony Hall is a dual qualified lawyer (12 years PQE English solicitor and admitted New York attorney) and majority owner and managing director of Pure Mint Recordings Ltd, an independent UK record label.

He has been appointed sole independent member of the International Legal Committee of the IFPI (since 2007), sits as an independent legal counsel member of the BPI Rights Committee (since 2005) and was an independent board member of the BPI from 2007 to 2009.

Anthony was admitted as a solicitor at renowned West End Media law firm specialists Schillings (Keith Schilling) in 1997 and admitted to the New York Bar in 2002. He worked as in-house counsel for Carlton Television (1998–2000) and Chrysalis Television (2000–2002) before embarking on setting up Pure Mint (www.pure-mint.com) in 2002 and establishing his own specialist media law consultancy (Entertainment Advice, www.entertainmentadvice.co.uk) in 2004.

Anthony is a musician and writer who released several records in the early 1990s and continues to DJ professionally today. He enjoys relaxing by reading, painting and composing at the piano.

PHILLIP R. GRAHAM (SENIOR VICE PRESIDENT, WRITER/PUBLISHER RELATIONS, BMI)

As Senior Vice President, Writer/Publisher Relations at BMI, Phillip R. Graham directs and oversees the writer/publisher activities in BMI's seven offices: New York, Los Angeles, Nashville, Atlanta, Miami, London and San Juan. Graham was most recently Vice President, European Writer/Publisher Relations, based in London. He now lives in New York.

In his prior position, Graham was responsible for providing a high level of service for members of the European music community who chose BMI as their US representation and for providing the same service to US writers whose works were performed there.

Graham moved to London in 1987 as Director, European Writer/Publisher Relations and was promoted to Vice President in 1991. He previously served for seven years in BMI's Nashville Writer/Publisher department.

A native of Evansville, IN, he received a Bachelor's degree in Business Administration from Vanderbilt University, Nashville.

RALPH SIMON (CEO, MOBILIUM INTERNATIONAL, WWW.MOBILIUM.COM)

Ralph Simon is acknowledged as one of the founders of the modern mobile entertainment industry. Over the past 15 years, he has been a prominent global mobile trailblazer and innovator, helping to grow the mobile entertainment and content industry, and playing a central role in its impact and presence worldwide.

Currently, he serves on the boards of several companies, including Hungama Digital Media Entertainment, India's major mobile content producer, and also mobile games maker Tunewiki.com. This summer, he will combine his passion for football (soccer) and mobiles to produce the mobile entertainment component of FIFA's 2010 World Cup in South Africa.

Simon heads the London-based Mobilium International Advisory Group, which provides high-level strategic advice and guidance to mobile handset makers, tele-communications operators, technology companies, media companies, movie studio and television networks, global music artists, advertising agency groups, brands and platform providers around the world. Specifically, Simon recommends unique, practical ways to expand a company's mobile business and achieve profitability by growing revenues and maximizing impact from the use and distribution of mobile entertainment content, mobile music, messaging, mobile media technology and applications.

As founder and Chairman Emeritus of the influential Mobile Entertainment Forum — Americas (MEF), the global voice of the international mobile entertain-ment industry — he works to raise industry standards and identify mobile revenue opportunities. Simon acts as a high-level advisor to a variety of entertainment and technology companies that understand the importance of producing cross-platform mobile entertainment content to reach the over 4.5 billion mobile subscribers worldwide.

He has been instrumental in pioneering the emerging field of Mobilology (previously known as Mociology), which examines the effects of mobile phone usage on psychology/behavior, sociology/community, culture, arts and entertain-ment, and the economy. In 2006, Simon established the world's first professorial Chair in Mociology at the DaVinci Institute for Technology Management in Johannesburg, South Africa, and today he supports an international endeavor to secure Mobilology as a bona fide field of academic study.

Prior to his leadership in the mobile entertainment industry, Simon co-founded the Zomba Group of music companies (and record label Jive Records) in London in the 1970s, building it into the music industry's leading independent music company. In the early 1990s, he came to the USA as Executive Vice President of Capitol Records and Blue Note Records in Hollywood and started EMI Music's global New Media division.

Simon correctly predicted in 1997 that mobile phones would become the indispensable voice/social networking-and-music companion for consumers and their increasingly mobile lifestyles. He persuaded US music publishers to embrace this new mobile medium by granting the very first ringtone rights. This spurred a whole new mobile entertainment industry internationally, and Simon was dub-bed 'father of the ringtone'. In the same year, he started the first ringtone company in the Americas, Europe, UK, Australia and Africa, called Yourmobile. In 2003, Vivendi Universal purchased the company and renamed it Moviso, and it was the leading mobile entertainment content aggregator in the USA for many years.

Between 2005 and 2008, Simon produced the mobile portion of the three highest profile concerts in recent global television history (prior to the

HopeforHaitiNow telethon in January 2010): Live 8, Al Gore's Live Earth and the TED Conference organization's Pangea Day. By integrating a mobile/SMS layer to all these events and capturing a new market using its preferred technology, Simon influenced the production of all future broadcast fundraisers.

For Live 8, he worked with the Live Aid organization, Sir Bob Geldof and Bono, to bring mobile connectivity to 12 countries simultaneously, delivering a television audience of over 700 million viewers. Simon successfully enlisted massive cross-platform participation in Al Gore's Live Earth global event, which spanned 10 countries and reached a global television audience of over 300 million viewers. That accomplishment led to his engagement as mobile producer for the 2007 TED Conference's prize-winning project, Pangea Day – a global telecast in 2008 featuring short films from filmmakers all over the world.

The influential trade publication *Mobile Entertainment* magazine has named him three times (2005, 2006, 2008) as one of the world's Top 50 Executives in Mobile Entertainment, and in 2007, he received its special award for Outstanding Contribution to the Global Mobile Entertainment Industry.

Simon works closely with the GSM Association, the global governing body of the mobile phone industry, providing expert advice on the future of mobile use and mobile entertainment. He also has served for the past three years on the Visionary Committee for MIDEMNet/MIDEM, the international record industry's annual convention that helps to shape the global music industry's best practices for use of intellectual property and music. He also works with media companies and artists to help shape their mobile strategies worldwide. One such example is RIM/BlackBerry's partnership with his client, Irish rock band U2, during their current world tour, which he conceived.

Simon is an internationally popular speaker on mobile and mobile entertainment. Over the past year, he has delivered keynote addresses at the important Music Matters conference in Hong Kong, the Canadian government's Canada 3.0 conference, Music Matters India in Mumbai and the World Copyright Summit in Washington, DC, USA.

At the GSM/Mobile World Congress in Barcelona 2010, Simon hosted and moderated the special Mobile Entertainment Summit. Other notable speaking engagements have included the GSM/Asia Congress (China), the India Telecoms conference (New Delhi), and major mobile conferences in the USA, Canada, Finland, France, Germany, Iceland, Israel, Norway, Spain, South Africa and the UK.

Ralph Simon is a Fellow of the Royal Society of Arts in the UK, where his 'Future of Media' lecture series is popular, and he is a member of the Academy of Recording Arts and Sciences in the USA. He is affiliated with the Mobile Life Sciences Forum in California, which is driving the early progress of the emerging mobile health industry.

ROD AARON GAMMONS (MANAGING DIRECTOR, APPLEWORLD DISTRIBUTION, WWW.RODAARONGAMMONS.COM)

Rod Aaron Gammons is a Director and co-owner of a group of highly successful businesses within the Planet Audio Group of Companies based at Pinewood Film Studios, specializing in the supply, resale, manufacture and distribution of high-end audio and film equipment to the creative industries. Planet Audio Group is one of the largest Apple computer resellers to the audio and video industries in the UK and Rod has been instrumental in the success of a brand new lighting product, Rotolight (www.Rotolight.com), which sold over 6500 units within the first six months of launch all over the world in 2010.

Rod holds a first class honors degree in Business from the University of Sheffield and gives an annual guest lecture at the Academy of Music on Corporate Social Responsibility in the 21st Century.

Rod has also written political research on other issues for David Cameron, now Prime Minister of the UK.

NIGEL ELDERTON (MANAGING DIRECTOR, PEERMUSIC UK)

Nigel has worked in the music publishing industry for over 34 years, beginning his career at Chappell Music in 1975. The following year he joined EMI Music Publishing, where he held various positions for the next 12 years, the last of which was as Director of Business Development, which saw the establishment of the first dedicated department for synchronization, licensing and promotion within the music publishing industry.

In 1988 he was invited to join MCA Music, heading up the synchronization department and liaison with catalogue writers, and a year later was hired by Lucian Grainge (currently Chairman and CEO of Universal Records) to join him at Polygram Music to concentrate on synchronization, general A&R duties and secondary exploitation of the Polygram catalogue.

In 1991 Nigel was offered the position of Managing Director of Peermusic UK, where he remains today. During his tenure at Peermusic, Nigel has been responsible for signing a number of hit songwriters, many of whom have gone on to international success with hits such as 'You Raise Me Up', which has topped the charts in both the UK and the USA, and which has now been recorded by over 250 artists worldwide.

Nigel is an active member of many of the music industry bodies. He holds, or has held, positions as Chairman of the MPA, Chairman of MCPS, Director of PRS for Music, Pop Publishers Committee (ex chairman), MCPS Commercial Committee, Trustee of the Performing Right Society Foundation (charity), PRS Deputy Chairman (retired December 2006) and Director of UK Music (formerly British Music Rights).

In 2005 he was awarded the Gold Badge of Distribution by the British Academy of Songwriters, Composers, and Authors (BASCA), and in 2007 he was invited to become a fellow of the Royal Society for the Arts (RSA).

MIKE MCNALLY (WWW.MCNALLYCONSULTING.CO.UK)

Mike McNally's music industry career started in retail with Our Price Records and then continued in the commercial departments of various media, including *The Observer*, *GQ*, *NME* and *Melody Maker* — where he was the Advertising Manager of the influential music weeklies in the early 1990s — before joining the launch team of the UK's first commercial national radio station, Richard Branson's Virgin Radio.

In 1996 he joined EMI Records. Mike worked across all of EMI's label operations including marketing, A&R, promotions and special projects for EMI Catalogue, EMI Liberty and Angel.

In 2005 Mike then joined the influential and innovative music and management company Nettwerk (owned by Terry McBride), where he ran the UK label operation and worked closely with the international management teams of artists including Billy Talent, The Stereophonics and Bombay Bicycle Club.

At the start of 2008 Mike launched his own consultancy, McNally Consulting. He specializes in developing marketing, promotion and branding strategies with UK and international independent artists, labels and digital start-ups.

Mike is also a member of the marketing committee of the Association of Independent Music (AIM), an AIM Mentor and is a board member of Un-Convention, a not-for-profit grass-roots music seminar for developing artists and labels.

In addition, he lectures on music industry topics at various universities including the Academy of Contemporary Music and the Bucks New University, as well as guest lecturing at the University of Mississippi and Belmont University, Nashville. Mike is also an External Examiner at Westminster University and is regularly invited to appear on, or to moderate, panels at industry events such as The Great Escape and Truck Festival.

EWAN GRANT (DEATH OR GLORY MUSIC, HTTP://WWW.DEATHORGLORYMUSIC.COM)

Ewan Grant, the founder of Death or Glory Music Ltd, was general manager at Ministry of Sound Recordings until 2003, at which time he created Death or Glory Music Ltd, where he signed, among others, Biffy Clyro. He then joined forces with Riot Management. Riot artists included Feeder and Joss Stone.

Grant has used his wealth of experience in management, recordings and music publishing to spot new talent and establish an independent music company with one goal: to find great songs. Alongside publishing, Ewan also consults for BMG Music in A&R and various blue-chip brands including Cancer Research. He also teaches at the Academy of Contemporary Music (ACM) and University of Westminster and currently manages Attack Attack.

OLIVER SUSSAT (WWW.SUSSAT.CO.UK/OLIVER)

Oliver Sussat is a freelance digital consultant and the founder of multiple digitally led businesses. Specializing in futurist strategies for product delivery and customer relationship management, Oliver's projects have been linked with multiple major broadcasters, publishers and record labels. He has a great deal of industry experience in the children's and youth sectors (with a focus on multiplatform brands) and also manages a roster of high-end television and music industry talent destined for those markets. Oliver teaches at the Academy of Contemporary Music and advises various media-based organizations on their digital strategies.

Glossary

A&R: The initials stand for artist and repertoire. Commonly describes a job role at a record label, but is also part of the function of music publishers in assisting songwriters and artists to develop, and developing the skills to evaluate potential music industry talent.

Administration deal: A legal agreement between the copyright owner and an entity to administrate an individual song(s) or an entire catalogue for a negotiated period of time. The administrator can be a publisher or a qualified professional (lawyer, accountant). Functions of administration include filing necessary documentation, copyright registrations, collecting royalties and negotiating third party licences (e.g. synchronization licences). Fees for administration are generally computed as a percentage of income ranging from 10% to 20%. You should get advice from an attorney before signing an administration deal.

Advance: A payment made to the copyright owner or writer before royalties have been earned. Advances can be made by publishers or performing rights societies and are recouped from future royalties. Advances are usually non-refundable.

Arranger: A person who takes a composition and decides which instrument will play what and where. This can be for a recording date or for a live performance. Aside from artistic duties, an arranger would arrange the music for the type and/or size of the band or orchestra that will be performing and/or recording it.

ASCAP: The American Society of Composers, Authors and Publishers, a performing rights society.

Assignment of copyright: The unconditional transfer of all rights contained in a copyright from the owner to another person or entity. This assignment generally concerns only the publisher's side of the copyright, although it is possible to assign the writer's rights as well. To be legal this must be in writing and should be reviewed by an attorney.

Audit clause: In publishing contracts, an audit clause gives the songwriter access to the publisher's or record company's books and records (usually once a year), so that the copyright owner can determine the accuracy of the publisher's accounting practices.

Blanket licence: A licence issued by a performing rights society it allows the music user, typically a television network or radio station, to play or perform all compositions covered under the licence without a limit on use, for one (usually annual) payment.

BMI: Broadcast Music, Inc., a performing rights society.

Brand: A name, term, sign, symbol, design, or a combination of any of these, used to uniquely identify a producer's goods and services and differentiate them from competitors.

British Academy of Songwriters and Composers: see http://www.britishacademy.com/

Business manager: A representative who helps the musician with financial planning, investment decisions, tax matters, monitoring of income from contracts, estate planning and other financial matters.

Business plan: The written document that details a proposed or an existing venture. It seeks to capture the vision, current status, expected needs, defined markets and projected results of the business. A business plan 'tells the entrepreneur's story' by describing the purpose, basis, reason for and future of the venture. It should be tailored to the reader i.e. investor, banker in its summing up and exit strategy.

Cash flow: An assessment and understanding of all cash coming into and flowing out of the business in a specific period. This can be based on projections or real time. Cash flow is the life blood of a business.

Catalogue: A collection of songs owned by a publisher or songwriter.

Commissioned work: When a writer or artist is engaged to write for a project. The rights retained by each party (including the company commissioning the work) are negotiable.

Compulsory mechanical licence: An exception to the copyright holder's exclusive rights of reproduction and distribution that allows anyone to record and distribute any commercially released, non-dramatic song as long as the mechanical licence rates established by copyright law are paid to the copyright owner of the song.

Controlled composition: Some artist contracts contain a controlled composition clause that applies to musical works written and 'controlled' by the artist. The record company considers these songs included on a CD a controlled composition, and per the clause pays the publisher of the artist's work a reduced rate, which is normally 75% of the statutory rate. The record company can also stipulate a maximum number of musical compositions (e.g. 10) on which it will pay mechanical royalties even if the CD contains more than that number of tracks.

Co-publishing: When two or more publishers share in the ownership of the copyright of the composition. In the USA it more commonly expresses the writer receiving his 100% of the writers share and also 50% of the publishers share i.e. a 75/25 split of the copyright.

Copyright: A bundle of exclusive rights granted by law to the creator of an original literary, artistic or other intellectual work, including songs and sound recordings.

Cross-collateralization: Allowing royalties from one song (or contract) to be applied against the un-recouped advance balance of another song (or contract).

Cue: An individual musical phrase intended to be used in a motion picture or television/radio show episode. The musical fragment may be part of a sequence of cues intended to segue without interruption between them. Cues used for underscoring can be used behind dialogue or to score visual action.

Cue sheet: A document containing a detailed listing of each piece of music used in a film or television production. The cue sheet lists the music by title, composer, publisher, timing and type of usage, and is usually prepared by the producer of the television program or film. A copy of the list should be requested by the Music Publisher and then filed with the performing rights societies.

Cut: When a song is recorded by an artist.

Differentiation: An approach to create a competitive advantage based on obtaining a significant value difference that customers will appreciate and be willing to pay for, and which ideally will increase their loyalty as a result.

Digital download: An encoded file of music (e.g. MP3, ACC, MP4) obtained via the internet from digital music stores, peer-to-peer file-sharing networks, and other media. The download may or may not have copyright protection (DRM).

Digital rights management (DRM): Methods used to control or restrict the use of digital media content on electronic devices, such as CDs. Also referred to as copy protection.

Diversification: A strategy to increase the variety of a business offering, a product or service to its customers. This technique is often used to reduce risk by having more than one area of income stream possibilities.

Economies of scale: The benefit achieved by larger production volumes allowing fixed costs to be spread over more units, lowering the average unit costs and offering a competitive price and margin advantage. Producing in large volume often generates economies of scale. The per-unit cost of something goes down with volume because vendors charge less per unit for larger orders, and often production techniques and facilities cost less per unit as volume increases. Fixed costs are spread over larger volume.

Ephemeral use: A use of music which can be used in media without a licence. It can occur in live television when the program will probably not air in future broadcasts. An example could be a marching band that plays a known song during half-time at a football stadium being shown on network television. It allows music to be used once in that context without a payment to the publisher or master rights holder. Ephemeral uses are still logged as if they were licensed and filed with the respective performing rights organization, but no payment is generated.

Exclusive songwriter agreement: A legal contract in which a songwriter assigns to a publisher a controlling interest to his or her songs written during the term of the contract, for a fee. Issues such as the term of the agreement, other songs to be included (back catalogue), reversion rights, the amount of the advance, and what percentage of royalty income the songwriter will receive are all negotiable points.

Exploit: To negotiate and secure sources of revenue for a song. In the context of music publishing it is not a derogatory term.

Exploitation: In publishing this means working to get your musical composition used: getting an artist to record your song or placing it in a television show, computer game, film or any other licence that provides an income stream.

Favored nations clause/Most favored nations (MFN) clause: In publishing, an MFN clause provides that an offer given to one party in an agreement is at least equal to the best offer negotiated with the other parties. An MFN clause may be used in negotiations for royalty rates (other than statutory) for a Greatest Hits CD. The clause would ensure that all publishers receive the same royalty rate equal to the highest rate negotiated. As an example, assume that licensor B initially agreed to grant a reduced rate, but under MFN. During negotiations, licensor A was granted a full mechanical. In this situation licensor B would receive a full rate because licensor A had received a full rate owing to the MFN clause.

Fiduciary duty: An obligation to provide a legal, ethical relationship to another party to carry out service or work. The fiduciary must not put their interests first, but act in the interests of the other party.

Grand rights: Also known as dramatic rights or dramatic performance, this term is used in connection with musicals, operas, ballets and other dramatic performances where the use of a musical composition is used to tell a story or as part of a story or plot. The copyright owner has the exclusive right to issue licences and collect fees for grand rights.

Harry Fox Agency: A company that represents music publishers in the negotiation of mechanical licences, synchronization licences and foreign licences, and the collection of music royalty income.

Hold: A period in which a song is taken off the market by the publisher and is not pitched because the song is being considered for use by an artist and a record label.

Infringement: Copyright infringement occurs when a person copies someone else's copyrighted items without permission.

Intellectual property (IP): A general term given to rights such as copyright, trademark, design and patent rights.

Library: Usually used in connection with production music companies. It denotes a collection of musical compositions that are licensed by the publisher or administrator for use as background, theme or score music, on radio, broadcast and cable television, films or video productions. The library is usually offered under a blanket licence.

Licence: To give specific permission. A permit from an authority to own or use something, do a particular thing, or carry on a trade.

Life cycle: A model depicting the sales volume cycle of a single product, brand, service or a class of products or services over time described in terms of the four phases of introduction, growth, maturity and decline.

Life of copyright: Life of the creator plus 75 years, or specific to the territory of that area of the industry.

Marketing mix: The activities controllable by the organization, including the product, service or idea offered, the manner in which the offering will be communicated to customers, the method for distributing or delivering the offering, and the price to be charged.

Master recording: An original recording from which copies can be made and which is at a standard to be commercially exploited.

MCPS: Mechanical Copyright Protection Society. See PRS for Music.

Mechanical licence: The licensing of copyrighted musical compositions requiring authorization from a music publisher or songwriter to record and distribute a song in a physical or digital format.

Mission statement: A statement that captures an organization's purpose, customer orientation and business philosophy.

Music publishing: The commercial exploitation of songs through the issuance of mechanical licences, synchronization licences, performing rights licences, print licences and other licences authorizing various uses of the songs.

Music supervisor: Someone who chooses and consults with the production company to acquire the music, songs and scoring for a film or television show. The director of the film usually has the final creative say on which piece of music is chosen.

Musical work: A melody and any accompanying lyrics; more commonly referred to as a musical composition or a song.

Net receipts: Gross income less specific costs determined by the contract.

Performing rights: The right to perform music in public. It is part of copyright law and demands payment to the music's composer/lyricist and publisher when a business uses music in a public performance. Examples of public performances are broadcast and cable television, radio, concerts, nightclubs and restaurants. When music is performed by a business they must obtain a licence to use that music and compensate the author (composer and lyricist) and publisher. Income from these licences is collected in the USA by one of three performing rights organizations: BMI, ASCAP and SESAC.

Performing Right licence: Authorization for the public performance of a song, frequently granted by a performing right society through a blanket licence.

PEST: Political, economic, social and technological analysis. PEST is a popular framework for situation analysis, looking trends. Analyzing these four factors can help to generate marketing ideas, product ideas, etc.

Pipeline income: Money that has been earned but has yet to be accounted.

Piracy: The illegal duplication and distribution of sound recordings. This includes unauthorized uploading of a copyrighted sound recording and making it available to the public, mix tapes, CD copying and selling material without authorization.

Primary research: Research and collection of data that do not already exist.

Print licence: Authorization from a music publisher or songwriter to reproduce and distribute a song in printed form.

PRS: Performing Right Society. PRS and MCPS are not-for-profit UK collecting societies that ensure composers, songwriters and publishers are paid royalties when their music is used, from live performance to TV and radio, CDs to DVDs, downloads, streams and everything in between. MCPS and PRS entered into an operational alliance in 1997 and in 2009 became known collectively as PRS for Music

Retention period: A period that commences immediately after the term of a contract, securing the control of the copyrights to the publisher.

Rolling advance: A further advance paid on recoupment of the first advance.

Secondary research: Research that has been carried out by a third party.

SESAC: Society of European Stage Authors and Composers. A performing rights organization.

Spotting session: A meeting between the composer and director/music supervisor to determine what type of music and where the music is placed in specific sections of a film. It is most often done when the film is at or near its final cut. Several spotting sessions may be done before the film is 'locked'.

Streaming: A one-way audio transmission over a data network. The audio data are not stored on the destination computer.

Subpublisher: A publisher in another country or territory retained by the original music publisher of a song to exploit the song in the foreign publisher's territory.

SWOT analysis: A formal framework of identifying and framing organizational growth opportunities. SWOT is an acronym for an organization's internal strengths and weaknesses and external opportunities and threats.

Synchronization: The placement of audio in a timed relation to a visual image.

Synchronization licence: Authorization granted by a music publisher or songwriter to use a song with visual images (e.g. in a film, television program, computer game or advertisement).

Term: The length of time a person or business is under contract.

Territory: Used in contracts to describe the range in countries of a contract.

Underscore: A score for a motion picture or television program (as opposed to the main title or end credits themes), usually used to mean that the music that is used under the dialogue.

Work for hire: As defined by the 1976 Copyright Act, a work for hire describes a composition that is prepared by an employee within the scope of their employment. The employer is considered the author and owner of the work. Even though the employee creates the copyrightable work they do not own it, and the employer treats the creator as if they did not even participate. Under this type of arrangement the writer will receive a fee for creating the copyright.

Index